POET ON
A SCOOTER

POET ON
A SCOOTER

by Harry Roskolenko

THE DIAL PRESS　　　1958　　　NEW YORK

Copyright 1958 by Harry Roskolenko
Library of Congress Catalog Number: 58-12000

DESIGNED BY WILLIAM R. MEINHARDT

A PATHFINDER BOOK REPRINT EDITION
Complete and Unabridged

Printed in the United States of America

ISBN: 979-8-8691-3890-3

Contents

Prologue	1
Chapter One OUT OF MYSELF AND MY COUNTRY, I GO	3
Chapter Two THE LAND OF THE COLLECTIVE UNCONSCIOUS	14
Chapter Three THE GLORY THAT WAS	32
Chapter Four SMOKESTACKS AND MINARETS	55
Chapter Five HOURIS AND OPIUM	102
Chapter Six THE DIVIDED LAND	172
Chapter Seven TOUCHABLES AND UNTOUCHABLES	180
Chapter Eight ANOTHER EDEN	205
Chapter Nine "TERRA AUSTRALIS"	219
Chapter Ten ISLANDS IN PARADISE	269
Chapter Eleven THE ROAD HOME	279

POET ON
A SCOOTER

PROLOGUE

TO THE MANY who helped to make some of the hazards of this odyssey less so, I offer my grateful appreciation, beginning with Andro Spolar, the Chicago building contractor, who ran interference for me in Yugoslavia; to Ilya the Greek, for an on-the-spot Olympiad during my first night in Greece; to Hank and Evelyn Malter, for the Istanbul interlude among the many Turkish delights; to the wolves near Mt. Ararat and the tiger in Ceylon, for letting me continue my journey; to the various American Military Field Teams in Turkey and Iran, for housing and feeding a reckless American citizen—specifically Colonel Nelson at Erzerum, Turkey, and Major Richie and Sergeant Bifora in Tabriz, Iran, who hauled me when the scooter broke down—and to their colleagues in khaki of Field Team #2 in Tabriz, for all the fun and games during the enterprising interim; to Colonel Resat Saran, in charge of the Mount Ararat garrison, for the spiritual vision at the foothills of Noah's mountain; to the businessman Mehmet Ali of Ankara, who almost made me his best man at a wedding I did not attend; to the Columbia University-schooled engineer, Fethi Ulgen, for an unforgettable fishing trip in the Kurdish mountains; to Mme. Pars of Teheran, who housed and fed me for very little; to Dick Snowden, the American vice-consul in Meshed, Iran, for his generous help when I bogged down during some of the more dangerous days of this bizarre journey; to Dr. Hovanissian of Nishapur, Iran, who saved my life; to an

English banker in Colombo, Ceylon, who accepted a check that later bounced; and to the many kind Italians, Yugoslavians, Greeks, Turks, Persians, Pakistanis, Indians, Ceylonese and Australians, my endless appreciation for seeing me through this odyssey on a scooter.

Also, to the manufacturers of the Vespa scooter, Piaggio & Company of Genoa, Italy, for the free use of their wasp-waisted vehicle; to Mobilgas of Australia and the United States, for some needed funds to buy oil, gas, bread, meat and wine; and to the many editors of magazines and newspapers around the world who bought my articles and photographs, giving me the wherewithal to do and see it all—my warmest thanks.

H.R.

Chapter 1

OUT OF MYSELF AND MY COUNTRY, I GO

OUT OF MYSELF and my country I went, to live in Paris among Left Bank bohemians, professional expatriates, and civil servants. I said a nostalgic good-by to New York, my apartment littered with an emotional ghost of a marriage gone; but Paris, with its exotic, Gallic flavors, soon gave me livelier ghosts to play with. I had lived there briefly through the maturing years, once when I was a rather young sailor of fourteen, no longer content to look at blue-green water or make grey ships my permanent, liquid exile. Paris was the exploding world for young poets. But it was just as electrifying for a man reaching middle-age, when added enchantments, as well as disenchantments, replaced the worn-out symbols of a marriage.

I left the day world of New York, and its compulsive pursuit of a better gadget civilization, to drift in the night world of Parisian cafes. There one relaxed and breathed; one also planned, worked hard and loved easily. I had a room on Boulevard Raspail, and each week I wrote several articles or stories for American magazines. I worked harder than I had worked in New York, when a twelve-hour day at the typewriter was the norm. But this was Paris, excessive in all things, but sensual and complete; part of me attuned, yet part of me missing. Here love came easily, over *apéritifs,* coffee and in tiny bedrooms.

Yet Paris was not my destination, just a stopover until the eyes and the heart undid the malices of a divorce, with its attendant disasters.

Four months after reaching Paris, I was suddenly adrift again, starting a bizarre, round-the-world journey on a Vespa motor scooter. My journey would create new tensions, excitements and values. But first there was a Parisian farewell party.

A United States Air Force pilot named Mooney, with whom I used to discuss old English music, gave over his apartment on rue Monsieur le Prince one Saturday night in May, 1956, and thirty old and new friends managed to bulk into the book-laden, record-strewn rooms. Translators, novelists, roués, poets, shapeful ladies, singers who sang and did not sing—and other successful and unsuccessful Parisian careerists—held up glasses to toast to my departure.

A blond Dutch girl, frail of morals but firm of breasts, drank too much and cried. Other guests merely ate and drank, to make up for a hungry winter. Some said sentimental things, for a friend was going away. Mooney, the host to all, whose Southern accent baffled every French cop trying to stop Mooney from killing himself as a speed demon in a sports car, talked about Purcell and Palestrina. "Then ancients really wrote it down, man. Take that Palestrina theme." Then Mooney changed his conversational attack, "Why man, you are intentionally mad. Go around that world out there on a little scooter! When I fly that SHAEF brass around . . ." and Mooney tacked again, taking a long drink from his commissary liquor, remarking this time with added vigor, "Stay in Paris and write another *Paris Poems,* 'cause otherwise you're crazy, fella."

In 1950 I had written *Paris Poems,* which Zao Wou-Ki, the enormously talented Chinese artist, had illustrated with six lithographs. It was Paris, the belly, brain, heart and eyes; and another deluxe edition was coming out. But if I was crazy, I didn't go to the couching specialists to find out; instead, I went to fifteen foreign consulates to get visas. I got linen Bartholomew maps of Europe, the Middle East and Asia, and traced the routes of Xerxes, Charlemagne and Genghis Khan. And despite the buds blossoming in the gardens of the Luxembourg, I was leaving the idyllic season. Mooney's musical evenings, and translators who had given up literary careers for the certainties of regular pay by translating other people's works for intellectual societies. I was assured by at least two of the thirty partying friends that at forty-eight, which is sort of middle-age, I had better settle down to some inconceivable civil-service permanence.

A young French photographer, who had spent a year riding buses around India, after his fifth Scotch and soda begged to go along. "I have done a book on India, going to all the places Buddha traveled to," he said. "I would however like to make many photos of Indian life in the villages. So you must take me along. I will get a scooter, too." But a drink later, the photographer of Buddha's travels was too far out of focus to remember.

"You're flabby, fella!" railed Mooney. "You're gonna tent out, ride in them monsoons and act as if you were twenty?" In my twenties, I had been an athlete, and I recalled, to bolster up my courage in face of Mooney's onslaught, that I had run against world-famous runners like Paavo Nurmi and Willy Ritola at Madison Square Garden—not that I had ever come close to winning. Later, I became a drawbridge operator, an occasional journalist in the labor movement, intellectually left, then the poet. And under five pseudonyms, I wrote a variety of novels for various publishers. I was the sitting-down man, content to grow physically flabby but no longer politically dangerous; for Marxism, under any other name, become an abhorrent social philosophy after fiftten years of living for the One World, One People, One Economic Salvation for All. I preferred liberal American capitalism, its revolutionary inventiveness, lots of worlds, various philosophies, and all sorts of people, especially in Paris.

It took a break in established routines to transform myself physically and mentally into the man envisioning the offbeat challenges of my private Baedeker. After the Paris party I packed two duffle bags and became a scootering gypsy, on the low roads of the world. I carried clothes for all sorts of weather, plus camping equipment, a typewriter and a Graflex camera, ready for everything in nature, woman and man. On May 18, I found myself in Genoa, the owner of a three-wheel scooter, with an Italian instructor, who spoke no English at all, yelling *"Pronto!"* and *"Piano!"* his teaching style and his vibrato-like antics soon persuading me either that he was a bad instructor or that I was about to become a bad driver.

The instructor assured me, further, that I would definitely be killed before I got out of his beautiful Genoa, adding to the fatalistic warnings of friends who had driven in Italy. I had never driven anything before, having once turned down the gift of a car, which my brother Mike offered me as a token for returning safely home from the war in the Pacific. Actually, I had a fondness for four-legged locomotion, disliking everything on four wheels. But since the scooter had but three wheels, my objection decreased, though it hardly increased my ability to drive this enlarged kiddy-car, which the scooter resembled. If this was mechanical madness, I was making the most of it, insisting that I at least be given a running chance not to be killed—and be driven out of Genoa by the instructor.

The next morning the instructor left me at the edge of Genoa, on the four-lane highway going to Serravalle. In back, engulfing me with memories of another world traveler who had gone toward the Indies via wind, sails, and wooden ships with iron men, was the blue-green Ligurian Sea. Bombay was only ten sailing days away, in an Italian liner; but many months away for me and over ten thousand miles. Soon enough I was in a maelstrom, with the passing Saturday traffic of cars, buses and motorcycles edging me off the road, clamoring when they came upon me struggling up the hill. My half hour's lesson had not made me an expert. I was unacquainted and afraid, all over the road, sailing into nowhere, frightened of stalling in this wild mechanical ocean of moving disorder, to which I had added my full measure of confusion.

I stopped abruptly, exhausted with tensions, not daring to look at the sea below, yet envying Columbus and the ease with which mere winds had carried men to new worlds. I stared at the extravagant olive trees and the vineyard-terraced hills above me, preferring to see peasants at the vines, their feet sturdily planted in nature. But I was off again, mechanically out of joint, jiggling gear shifts that did not respond like ankles and hand reflexes. The machine lunged ahead, back into the mechanical madness that I had just fled from . . . and it stalled. I was islanded, embarrassed and lost in the dead center of the traffic, in a strange world of honking humans, their faces motorized and their mouths geared to screeches and exclamations all aimed for me.

I thought of taking my camera out and photographing the hundreds of vehicles I was holding up on the hill turn, a line that likely descended into Genoa; but then I saw the screaming faces, the comedy and the laughter that would follow the antics of this strange American who was not content with Purgatory, but wanted all of Hell, as well.

I fled by getting out and pushing the scooter to the side, reflecting all the more on donkeys and mules, whom I now loved like an old muleteer. Had anybody gone to India or around the world on a mule? I could be the first one, even if it took a few years. I would get to know the mule humanly. Instead, I was off again, becoming Columbus' iron man, hugging the curb, going from six miles an hour, which was my first speed, to ten, which was my second. For an hour I moved in this fashion, frightened of shifting into high gear, when I could go thirty, forty, and even fifty miles an hour. I had dreams of the third gear, as some saintly thing that came with the proper proportions of goodly works done on this earth . . . and I wondered, debit and credit, as I climbed through the Ligurian Apennines how saintly or sinful I had been in Paris . . . whether my lack of rewards was this built-in nightmare. When I dared shift to third, I saw myself passing a beloved trotting mule pulling a small cart loaded with hay. It was a great, magnificent experience: I had passed a vehicle. I saluted the driver, but mostly his mule, in a language I invented at the moment, though I just managed to miss hitting the mule as I scootered by.

With near hits, many misses, and halts, I turned myself into an expert by the time I reached Cremona, 120 miles from Genoa. Once, when the scooter would not start, a well-dressed fellow traveler on a Lambretta scooter paused in good fellowship to help find the reason. An hour later he found the reason, his suit neatly smeared with oil and grease, his right shoe torn from getting caught on my starter, but his humor still intact. Mechanically repaired, I broke out a bottle of Chianti and we toasted his embattled suit, though I tried to pay for cleaning it. We also toasted the New Italian Renaissance he convinced me was on its way, beginning first with the films, then with the opera, to be followed by a new form of political morality; and after that—the Renaissance theater, without Tennessee Williams or other foreign imports. The television world had already invaded Italy, and the night before I had watched whole families of Genoese at a bar voting for the best television play of the year. When I mentioned this collective voting, the now-not-so-well-dressed gentleman scorned my enthusiasm for the event by saying, "They will look at anything so long as it's free—but to pay, that is something else. The opera in Italy must run on a big subsidy; for Italy without opera would not be Italy."

But Italy was opera, with its natural subsidy in dark alleys, tenements, green villages and the lovely large women. And since I had never loved the slenderized ideal American, who nevertheless believes in the bulging breast and the heaving bosom, I found the Italian ladies fully appealing, dramatically lusty, impish and courteous, their charms as complete as an Italian dinner with Orvietto to drench one's taste buds. At the Alberghi Brescia where I stayed the first night, Madame la Patrona was fat and engaging, humorously typing out the hated form that every foreigner must fill out, much as if he were a Chicago gangster suddenly deported to his native village in Italy. But Cremona was much more famous for its great violins than for exportable gangsters. The people sang and did not shoot, though they also specialized in politics, too much *vino,* and spaghetti spiced with garlic and garlands of rhetoric.

I was vibratingly happy, though my body felt like a motorized eggbeater as I haltingly and shakingly went to sleep. I had not been killed, though I had come close to major mishaps; and this was my first day and my first night, with a radio playing *La Boheme* lulling me to sleep's green world.

Madame la Patrona eyed my American Express check the next morning, then slowly undid her right stocking high above a fat knee, unwinding from its padded depths an American one-dollar bill. "Is this the same?" she asked.

"It's ten times more. President Eisenhower himself is behind it," I answered.

As I packed, she examined the scooter, then asked, "Who make the APE?"

I pointed to all of Italy on my Bartholomew map, then found Genoa, which had a red circle around it, and said, "Signora, have you no shame? This is the Italian motorized-renaissance; so a gentleman said high on the Apennines. The Vespa APE model is all-Italian including the name, which means 'bee.' "

"Madonna mia!"

Mantua, the fortress city of Vergil, came up suddenly, its thirteenth-century buildings and St. Andrea, the church designed by Alberti, shaping themselves within the heat haze that rolled over the Lombardy mountains and upland pastures. The Mincio river, breaking into three inlets, almost surrounded the city, and like liquid passages into time, reminded me that two thousand years ago Virgil walked these streets thinking Epicurean philosophy. The first impulses of the *Aeneid* rose here; and Augustus Caesar, once a fellow student of Vergil's in a Roman private school, had been his sponsor. But Mantua was giving unto Caesar what was his—for a violent political struggle was going on all over Italy, with political parties outdoing each other in the violence of their satire and political promises. But there was broad Italian comedy, with *opéra bouffe* humor in some of the political satire pasted on the stone walls of Mantua. One poster showed Bulganin and Khrushchev in the nude, each being posterior-injected with brainwashing fluid to rid themselves of secret or open residues of the Stalinism they were now disavowing.

For lunch I had cheese, bread, salami and red wine as I sat above the Italian scenery, suddenly aware that I was a free citizen of the world; that I had really embarked and was en route, and was no longer making intimate speeches to friends about my journey to be. The sense of action released the pent-up aching rivers that Whitman had foreseen in the modern American male; and as I passed Montseris, like a Bellini painting with its deep greens and reds and Chinese serrated hills, I knew that I had not lost my need to go back into time. I was entering the lush geography to discover the vaporish worlds that had been only vaporish images in poems. I was eating salami in a medieval setting, my New York malices evaporating as I approached Montagnana, another fortress of the past, enshrouded in Latin goodwill.

I camped at Chalet Helda in Padua, a privately run camping ground filled with Germans lugging cameras and *Gemutlichkeit*. Shakespeare had been here, dramatically; for Padua had been a setting for one of great William's many murders; but Padua and Camp Helda, with yodeling campers in leather pants getting Mozartian after midnight, were like Tyrolean versions of Times Square, their noisy insistence on Germanic ways of life and love awakening me everytime I found my Elizabethan sleep murdered. Ah, for the noises of nature and the silence of man!

Camp Helda's mock heroism made me the early voyager, for I was scootering at 6:00 A.M. toward Venice through the Italian mist, going forty miles an hour down the Autostrada. When the engine died suddenly, I discovered that all campers, unlike all boy scouts, are not honorable; for most of my gas had been siphoned out of my tank. A passing Vespa scooterer offered me a liter of gas, providing I elected myself as chief sucker of Ye Mobilo Gasso. He handed me a thin rubber hose, and in I sucked, twelve times before a liter passed from his tank into mine. I was drunk with fumes and I had to go back to the *vino* to clear my throat and my head, to keep myself from losing my sense of direction. The combination, however, almost air-lifted me, and I became a racing driver, passing hundreds of mules, trucks and cars, in a mad dash for Venice and the Lady on the Venetian waters.

Venice must inspire retired tugboat captains as they tour its sunken grandeur and liquifying landscapes. I imagined them going down the Grand Canal in Diesel tugs, towing sail-less caravels loaded with bullion, three golden balls swinging from the booms, the same golden balls which Cosimo de Medici had used in his fifteenth century banking-pawnbroking empire. All things were related in my mind, church and man, man and money, money and the Florentine financial empire that once had so much of Europe in hock. But Venice today was tourism on tour, with 3,000 pictures taken every minute and 35,000 cars parked every few hours at the AGIP parking lot. The modern Medicis had their latter-day commercialism, balls and all.

I took the ferry boat to the Lido, which looked eerie from the sea view; and I gaped with tourist eyes at Milady Venice's best-dressed exhibition, very watery above her knees, splendid of bosom, high of breath, higher in commercial instincts, and low when the picture snapping is done and the heart looks out at all that once was and is no more; much like an ancient Madame who knows all the tricks but can't find a client to go to bed with.

At St. Mark's I sat over an *aperitivo Americano* at the Piazza, amid the fluting, fluttering pigeons, talking with a Harlem Negro fascinated with the view. "It's so different," he kept on exclaiming, his voice rising with possibilities. "Imagine, if we had all this along the Harlem River. Man, what a city we'd have in little old New York—what a great city!" I agreed, if for other associated reasons; thinking that perhaps the Giants and Dodgers would then stay home and not leave Harlem and Brooklyn to wander afield as a ball club and help New York disappear as the capital of sports and/or spiritual values.

But baseball is not a religion for all despite its almost sacrosanct similarities and ritual-like enthusiasms; for while sitting alongside the walls of St. Mark's cathedral, eating bread and drinking wine, I met a young Italian executive of a travel bureau, who thought my dirty khakis made me a traveler, if not an American. He offered me his Ritz crackers, then remarked in studious English, "Sir, this is the first time that I have been to Venice. I have been to Rome twice, to London three times, to Paris five times, and here but once. It is so beautiful, Venice; so very beautiful," and his eyes took on an extraordinary distant worship as he added, with religious fervor, "But ah, the Empire State Building, that is indeed great, sir. Italy, and especially Florence, where I live, is too much with antiquity . . . just like death. The Empire State lives now."

Having seen Venice, unconquered, I was soon deep in the Italian roads, tooting my way toward Trieste; for the roads out of Venice were crowded with speeding adventurers. When I honked at a brightly dressed lady pedestrian, refusing to knock her down, she finally moved her large lovely rump off the road and exclaimed in refined British English, "And I thought it was a cow. What a lovely little wagon!" I saluted, bowed graciously, remembering Browning's "Pippa Passes"; though she was, in her over-all British accents, much larger than Pippa.

Photographing and noting, a contemporary Gypsy come to the earth on wheels, I traveled over one hundred and fifty miles a day, taking in the splendor and the sadness of pastoral Italy. For if the land is to give bountifully to the peasants, they must at least possess that land. But Italy, radiant in May, gave only promises.

Nearing Trieste, I smelled the broad reaches of the Adriatic spewing up along the frontiers dividing Italy from Yugoslavia. I became overly aware of the word "proletarian," the philosophical meaning of "dialectical materialism" in all its associations. A few miles away, in Yugoslavia, Tito was trying a new approach to the problem of socialism to fabricate a happier breed of man. When I camped that night at the Trieste camping grounds, my Danish, German and Austrian campmates, some of the latter from Chalet Helda, in Padua, I added to my knowledge of the happy Danes, the gloomy Germans, the sad Austrians, all trying, by musical strumming and furtive fury, to be at one with the heroes of the Niebelungen myths. I escaped, refusing to cook, taking a Danish girl camper out to dinner; and we found a nearby restaurant and ate veal with a slight tang of time, soggy potatoes along with sour wine. Depressed with the fading food, we danced to a jukebox American folk song with a South Carolina twang echoing in our wake. The owner of the resounding voice was an Air Force sergeant, with an Italian girl in tow who worked at his nearby base—twenty-seven miles away. "We come here regularly," he said joyfully. "The food's good and this place is so damned American, fella."

I left, with the Danish girl becoming sick; for if the veal was not an Italian antique, it was aged enough. She vomited cheerfully, with many apologies for being so human. After that she felt hungry and demanded smorgasbord. "I have nice things to eat at the camp. You buy a bottle of good wine and we will dine properly," she said.

We dined, near a little stream that ran by the camp, She was bound for Holland in her Volkswagen and our paths divided by the morning, when the hills of Yugoslavia emerged from the haze.

The totalitarian vista had a peculiar impact on me, for this was the first time I had come so close to the geography of communism. The Adriatic Sea made a more entrancing background; and though I had a three-day visa and was an acceptable tourist for Tito's dialectical domains, I hesitated about entering Yugoslavia; but it was on my map and the way to Greece.

The entrance to Trieste had an intense charm, with blue gardens and white houses splashing above the mountain road. But Trieste had been lost and found, having been everything to all of its many conquerors. When I saw *Buvete Coca Cola* my instincts for things not overly American rioted briefly; for it was splashed over a lovely stone wall enlacing a blue-green garden. Objecting vigorously, I soon changed the sign to read *Buvete Roskolenko Vino*.

Chapter 2

THE LAND OF THE COLLECTIVE UNCONSCIOUS

THERE ARE various ways of entering Yugoslavia and the best road in Yugoslavia is the Autoput going to Belgrade; but along the Adriatic, where the scenery is infinitely better, the road is bad, though it cuts the distance to Greece by 240 miles. I decided to go via Belgrade, and into the social heart of Yugoslavian communism, preferring to increase my facts about Tito's contribution to the welfare state rather than have more photographs of the Adriatic's charming islands hugging the brown coast.

But though there were various ways of entering the confined country, by noon I had tried four approaches to the land of the Collective Unconscious, circa 1956. Eventually I managed to convince the Italian frontier police that it was Yugoslavia I was going to, not Austria. I was redirected through devious mountain routes to a place called Farnetti, where the Yugoslavian border guards looked like Pennsylvania miners coming up from the pits, their daily proletarian garb worn with heroic indifference. They were officiously polite yet likable, though I expected a lecture on the virtues of Marxism to go along with the rubber stamp that finally made me a Yugoslavian tourist. But when I requested an extension to my three-day *en passage* visa, the custom's officer was quite unconcerned, adding humorously, "In Belgrade you can get anything, including a prison sentence for spying."

I sped off, going past stony fields, over harsh mountains on untraveled gravel roads, feeling excited and almost tearful as I saluted Slavic men and women walking with hoes and scythes over their shoulders, bound for the fields. I felt my own Slavic affinities, my Russian parents having bequeathed to me my Slavic features and coloring, though I resented this heritage at times, especially the morbid strain, the religious fanaticism, the lower depths of the heart, but not the Slavic temper and energy that had also been included in their varied gifts to me.

Yugoslavia is a land of magnificent scenery and poor-but-magnificent people. Soft hills and valleys soon replaced the stony fields, and lush squares of greenery dominated the earth like mammoth Grandma Moses paintings. The children, dark eyed and badly dressed, had a quiet dignity in their faces. Yet everything came off either black or white, exaggerated and totalitarianized, and I heard spelled-out objections to the regime whenever I looked sympathetic enough. Remembering the customs officer's comment, "You can get anything, including a prison sentence for spying," I refrained from making obvious leads to better stories, allowing the worker or the peasant to talk on at his own rate of political exposure.

Nearing Postumia, two boys of seven waved nervously, their schoolbooks dangling in universal style. My two hitchhikers soon were sitting on my baggage, chattering happily, eating my Italian candy, answering *Da* to everything I asked in German, and *Niet* when the ran out of *Da's* as I attempted to discover the human ABC's of Tito's land. They grinned, patting each other, making soft exclamations of joy; for this was their first ride on a motorized vehicle, and the scooter was more exciting than a motorcycle or a rickety wagon. After three miles they pointed to a little white schoolhouse, where their schoolmates were playing in the yard. I stopped, they bowed, said thanks; and as I went off, they yelled their gratitude with abandon.

The restaurants that I stopped at were more like Bowery handouts attached to flophouses; a broken-down, spiritless ingathering of worn-out furnishings, frayed curtains and broken tables; the utensils, discolored with years of use, were brown, chewed on and dangerous; the cups more broken than whole; and the proprietors, who were very tiny capitalists, looked brutalized in their barren cafés. People ordered potato soup and ate huge lumps of black bread. It was a diet that fattened them for robot-like activities. They did not smile, they did not laugh as they groaned their reactions to the political blueprint that was contemporary Yugoslavia. At one restaurant, a fat spitz sat underneath my table looking even sadder. When I threw a lump of bread to the dog, I thought I heard it whimper, "Thanks, comrade."

I expected to see proletarian legions of labor moving over the landscape, but I saw little sweat until I reached Ljubljana, forty-five miles from Trieste, which I reached late in the afternoon. Once I strayed and got lost. But a peasant set me right again, though I, by past association, asked for Lubyanka, the famous Soviet prison, instead of Ljubljana. Not that he noticed my error; for the prison he was in had an open sky, but its barriers locked up the land and its people.

Road conscious, knowing the pitfalls of tiny wheels and a scooter's jarring imbalance on pitted roads, I was soon balancing over jagged, ragged tracks toward Zagreb, sluicing over hair-pin turns suited for neither man, beast nor scooter. I had miscalculated the distance, and it was almost dark when I reached Nova Mesta, still some distance from Zagreb. But at the Hotel Metropole, which is a universal name for bad hotels, the manager suggested that I pitch my tent in the back of the hotel, as he had no room. He had no room for another American traveler, as well; for standing at the desk, deep in Yugoslavian speech, was a fat, whimsical capitalist who called himself Andro Spolar. He beamed capitalism, without the usual cigar of the Leftwing cartoons; but he was critical and cynical about Yugoslavian communism, rattling the poor manager who had turned both of us down to tent the night in Grandma Moses's pictorial heartland.

"There is a hotel fifteen kilometers from here, at Dolenske Toplice. Are you an Italian?" asked Spolar, looking at my Italian license plate.

"American, but my parents were Russians."

"Then we are both cosmopoliton. I come from Chicago, but I was born near here. Let us go to Toplice, comrade."

Outside was Spolar's 1956 Buick with an orange-colored Florida license plate. I followed him for sixty minutes as he darted up and down dirt roads in the Yugoslavian blackness, during which time he picked up five peasants and went out of his way to get them home. When we finally arrived at Dolenske Toplice, a famous hot spring, Spolar had given away a lot of American goodwill in his capitalistic Buick. His conversation, he told me later, was strictly on how to make money and how to become an entrepreneur. He was a retired Florida and Chicago building contractor, and he knew many of the racketeers who had muscled into the building trade unions. He likened some of our prize gunmen to Tito, but always with enough comedy to make his listeners think that everything that happened in the world was just a lot of vulgar comedy; that is, everything but making money, which Spolar proclaimed was the only decent pursuit for all proletarians, especially if they believed in the dialetics of value, price and profit.

I believed in a hot bath at the hot springs, and into the tiled pool we went, along with very fat, not pretty lady guests who plowed their way about the large pool and exclaimed, *"Ah, Das ist gut,"* and I, taking a cue from Spolar, said that all was good in this best of all capitalistic worlds.

Local and visiting bigwig communists were to descend on us later, to splash their way toward a great occasion being readied for the next day: a firemen's holiday celebrating the one-hundredth anniversary of the founding of the Dolenske Toplice fire department. The restaurant was gala with bunting, the dance hall with more of the same, and all the paying and nonpaying guests looked clean, well cooked from the hot baths, waiting for the bands to strike up the marches.

A crowd gathered around my scooter to admire its lines, as if it were a Mestrovic sculpture. One, more avid with economic determinism, finally asked the price. When told it was just a few hundred dollars, he shook his head, adding, "With my wages it would take me fifteen years to buy one. America is very good to the workers. Are you a capitalist like Spolar, or a worker?"

"I am self-employed, so I must be a bit of both. Ah, America!" I answered, going to the dining room where we were seated with servile ceremony by a waiter who imagined we were visiting commissars from another country. The dinner was abundant, for this was a rest resort, with underspiced food served along with overflavored speeches. We had German *Kalbbrust,* Soviet *shashlik,* Japanese cucumber salad, Austrian pastry swimming in heavy cream, and a liter of white Yugoslavian wine, as well as several international *apéritifs* in advance, all of which came to one dollar for Spolar and me, with our hesitant tipping not exactly frowned down. It was a Marxist appreciation for the bounty of the capitalistic Amerikanski abroad.

Dolenske Toplice, Slovenia, was also playing the host to returning Yugoslavians who had lived as workers in the United States. One of them, now the proprietor of an inn, had left Cleveland during the 1929 depression. Over coffee, he said, sadly, "I must have guessed wrong, for I should have stayed on in Ohio. Now all I have is a Chic Sales toilet, a café that sells ersatz coffee, and no customers but two Americans. But they are building a modern highway here, so that is at least like the United States. Ah, how beautiful Cleveland must be now!" When I asked for cake to go along with the bad coffee, he shrugged it off, replying, "They have cake in Belgrade and in the resort restaurant for the commissars. I only have black bread. Do you want some?"

Later, needing gasoline for the scooter, I asked how much a gallon cost. Another shrug, but much more vigorous this time; and in a voice that was hardly restrained, he said angrily, "I wouldn't know. Who can use it around here?"

Another returning American of Yugoslavian extraction was sixty-seven-year-old Frank Wolf, who had left Yugoslavia at sixteen, worked in a paint factory in West Virginia for his first fifteen years as an American, then in a Pennsylvania coal mine for the next thirty-five years, retiring in 1955 on Social Security. He received a monthly check for $97.50, which he exchanged for 60,000 dinars, or twice the normal rate, with the government anxious for his dollars. He had recently married a woman of thirty-three, who gave him a child and heir to the piece of land where Tito let Wolf grow grapes, which he turned into wine. Though hanging on to his American citizenship, Wolf nevertheless decided to send back his return ticket, saying, "I think that I will die here, not in Pennsylvania."

"You will die from starvation or from too much wine," added Spolar. "Here, have some Eisenhower chocolates," and he pushed a pound box into Wolf's hands. "Better, feed it to the kid, so he can grow up strong and capitalistic."

"Aren't you at all worried, talking the way you do?" I asked Spolar, noticing that Wolf was embarrassed.

"Tito won't touch me. I worked in the mills with his father; also, I had lunch with him once. So!" and Spolar roared with laughter. "I even gave him some Eisenhower sweets; but he's used to that from Point Four, or whatever they call American aid."

The next day was the firemen's holiday, with a touch of old home week as politicians from nearby Ljubljana, in stock blue suits, vied with the firemen, the martial music of the brass bands, the parading schoolchildren and the local peasants, all praising the shy firemen and their fire-fighting ancestors for keeping Dolenske Toplice from burning to the ground. The partisan cavalry on farm horses wedged up front, circling the dais, so that the rumps of the horses became all the more significant. When I tried to take photographs, I was only allowed the proximity of their rumps, all else, including their forelegs, being agricultural military secrets. I managed, nevertheless, to get the commissars on the dais, the children running back and forth between the speeches, the male and female athletes who performed gymnastics, as well as several officious policemen, who were happy to get into the picture, striking poses, 1880 style, before a visiting communist official decided to confiscate my camera. In a few minutes I was in the police station and under examination by a young police functionary, who finally let me go with the warning, "This is forbidden, comrade!" My film was still intact. And when I developed the proceedings of Dolenske Toplice in the quiet of Athens, I had a fine collection of Yugoslavian horses' arses in my album of communized memories.

Spolar went to bat for me at the police station, his Buick hardly a good talking point for either of us; nor the joke that he told the functionary. "Do you know why it is considered bad taste to clean your teeth here before eating?"

"No, but what has this to do with taking pictures? Why is it bad taste, comrade Spolar?" asked the annoyed functionary.

"Because then you don't know what you missed, comrade policeman!"

I missed going to jail, I was assured by the straight-faced functionary, as he waved us out of the police station; I also missed a lot of speeches about the greater production expected from industry; but Spolar made up for my losses with more jokes. Spolar had taken four thousand lead pencils with him to Yugoslavia, and on each pencil, in red, was printed, "Andro Spolar, Chicago Capitalist, U.S.A." These he gave to every child he passed between Trieste and Zagreb; and he illustrated a hundred ways the children had of waving their joyous thanks for the pencil. Without doubt, as Lenin had once said, he who has the children has the State, eventually; but Spolar had them, at least for a little minute, with his little engraved legend about himself on a proletarian lead pencil.

I planned to leave for Zagreb the next morning, having gained three pounds during the starchy, hot springs interlude. At breakfast, Spolar objected to my Russian rush, as he called it, promising to get my visa extended, "even if I have to go to Tito myself. Also, take a look at the sky!"

A cloudburst came tumbling down. An hour later it cleared again, and I was off to the hills, entering an archway of black clouds that almost touched my head. I bumbled over the bad road for thirty miles, at which point the famed Autoput was to raise its lovely stretches into Belgrade. At Zapresic my scooter was almost disassembled by a swarm of little boys, who moved from the unmovable to the movable parts, feeling each piece as if it were a Christmas toy suddenly descended from the black clouds. When the scooter refused to start—not that any parts had actually been removed, just lovingly touched—the boys, in a collective crowd of chattering energy, happily pushed me up a small hill until the motor kicked over. But they were not satisfied and insisted on running with me. To please them, I gave them a lift, taking three at a time up and down the looping road; then I went back for the next load. After a half-hour of play, I started away with my capitalistic scooter, which I could not give away as Spolar could his 4,000 Chicago, capitalistic, U.S.A. pencils.

But the roads got much worse by the time I reached Zagreb, a dull, prison-like, gray city. I was convinced that I was not in the automotive age, at least not in the scooter stage of it. But outside of Zagreb the Autoput loomed up. I passed new housing blocks in the fields to my left; boxlike squat structures aching with poverty and architectural dullness—and these were the tenements housing the tattered tenants of Tito's socialism.

But the pastoral setting was always amazingly beautiful, no matter what else was indecent, like the many factories named after Tito in a Stalinist habit never really changed. Cowherders, sheepherders and horseherders cropped up everywhere to humanize the land. The women watched three cows each and the men looked after three horses. The children played with three goats, representing their collectivized or private stock, each with a stick to lean on or for throwing at the animals when they grazed onto the cultivated land.

My scooter stopped dead, running out of gas just as I reached the base of a hill. Behind me were ten large army trucks filled with Yugoslavian soldiers out on maneuvers. Since it took a few minutes to refill from my spare gallon-can, I laughed openly; for I was holding up a small segment of the Red Army, a scootering diversionist in a new form of roadblocking, temporarily disorganizing two hundred laughing soldiers who offered to lug me to Greece, if I could get them over the frontier.

At the Novska Servisna Stancia (auto service), a mechanic, who had once driven ex-Ambassador George Allen's car, proceeded to disorganize the scooter's internal system. After watching him for an hour, despairing at the mechanic's furious fumbling, I asked him to discover oil elsewhere and invited him to have a drink of schnapps with me, realizing it would be less costly if *he* did not do what the Vespa handbook insisted must be done after the first 1,000 kilometers—to drain the crankcase and the master brake cylinder.

Happily, he dropped his search for oil and accepted my invitation for schnapps, during which time he told me that he had fought for the British at El Alamein; that he admired George Allen; that he spoke seven languages; and that he really preferred lemonade to schnapps. As a mechanic he received 10,000 dinars a month, paid out 800 dinars for rent, and 3,000 for food. He slept in a small room with four other mechanics and worked seven days a week. His only relaxation, besides playing with a small fox cub, came when he found American or English travelers, whom he regaled with his memories of El Alamein when he was not voicing a not-too-peculiar desire, as when he said, "I hope one day to find an adventurous-enough-Englishman who will give me a lift and get me over the border by hiding me in the trunk case. Of course, if they found me, I would be better off at El Alamein, slightly shot up." I offered him a ride to Greece and to freedom; but he shook his head, remarking philosophically, "By the time you got to Greece with me, it would have been reported to all the police and then what?"

"You insist on sticking up for your rights, like a good Yugoslavian proletarian. Then you run for your life!"

But I ran, having lost too much time, with my crankcase still undrained and gritty. At Slavenska Brod, not too far away, I was sure that I would find a mechanic who knew the mechanics of the scooter as well as the problems of life; for Slavenska Brod, with its motel-like *servisna,* had a Scotch engineer as a paying guest, on his way home from Baghdad. Over schnapps, and after draining the oil from my scooter, the burring engineer said, "I had to pass myself off as a German in Athens because of the anti-British feeling over Cyprus. I know exactly one word in

German—*'Dankeshön,'* which I used on all occasion. But when a talented linguist-clerk at my hotel greeted me one day by saying, 'And how is the wee Mr. Rob-ber-r-rt Bur-r-r-ns today?' I answered in my best bur-r-r-r, 'Never r-r-read the old gent any mor-r-r-re; but *dankeschön.*' I've been nursing my old Morris car along, hoping to make the channel ports with it. Mon, but it's in a ver-r-ry fr-r-rail shape; and the r-r-r-roads here are well bur-r-r-red, at that."

He offered me his map, neatly marked with his past travels. "This may help to get you through. Notice all the lines in red? Well, they aren't exactly r-r-roads. They are simply r-r-red lines. and you can't travel very far-r-r-r on a red line. And in that things of yours, Mon, how far-r-r-r will you r-r-r-really get? Why, I met a young Frenchman who had come all the way from Saigon in a scooter, and he'd been three months on the r-r-road; but he quit in Beirut. He was not a motorized Ulysses, I fear-r-r-r."

A cripple in a manually operated wheelchair pulled up just then. The attendant helped him out of the chair and carried him to a room in the motel. The cripple was on his way to Ljubljana, having left Belgrade several weeks before. He had massive shoulders, powerful hands, a thin, dignified face, but his legs were badly mangled. Refusing to accept government help, he was on his own, heading for home, with all sorts of bureaucratic repercussions following in his wake. I said to myself "There really goes a man."

The attendant, returning, told me how poor everybody was in the Slavenska Brod region; that he had but two shirts, both of them worn to a frazzle, and could I perhaps sell him or give him a shirt? I gave him a blue shirt I had bought in Paris, happy to unload some of my colorful middle-class holdings.

When my speedometer finally showed that I had traveled one thousand kilometers, I celebrated myself, toasting with a schnapps at the next village. I had been a week on the way; my body was hard, my color a deep pink-brown, my hands laced with blisters from the jumping handle bar shooting up at me from every rut and pothole. Months away, and thousands of miles in some endless vista, was India, lost in a mystical geography of place and people. But the distance registered on my speedometer was not mystical, showing with mathematical precision my first week's adventuring amid stranger political systems. I was the happy traveler, completely carefree in a world that had only worries. Whatever turned up now was only too human, I knew; for this was the best of all possible worlds.

At Froska-Gora, I entered a series of savannahs and inlets, with rain forests running on both sides of the roads, and wide valleys meeting the forest at varied intervals. I was entranced by the bamboo-like trees that filled out the dripping forest, where woodsmen in their little boats rowed, marking off those ready for cutting down. The scene was as exotic as a poem by Hart Crane, or by Rimbaud, who must have imagined something like this when he saw his great pageant and visions in *Le Bateau Ivre*. When I asked a rowing woodsman if there was any wild life in the woods, he answered smilingly, "None but I." Scraggy and bepatched, he was a cross between Senator Kefauver and Wallace Beery, sans coonskin hat, but filled with homilies and grog for all. In a Beery-like gesture he offered me a swig from a skin flask. Expecting wine, I drank deeply. Disappointed at my expression but amused nevertheless, he said, puckishly, "We have no money for wine, so we drink the next best—water."

After Froska-Gora, I encountered more animals than humans or vegetables. A pig crossed the road, stood transfixed in the center, waiting to be roasted by my scooter's hot wheels; but I swerved and did not have pig for dinner that night. A few miles away some geese did the same, waddling slowly but authoritarianly down the middle of the road. I also did not have goose that night. I had more *Kalbrust* and black bread.

I passed a stalled car with an Arabic license plate; a tall thin man was leaning over the motor. I yelled back in English, "Can I help?" But when he said, "Don't bother, old chappie," I, believing this to be a British understatement, bothered and returned to the obvious Englishman who had come all the way from the Kuwait oil fields and who, like the Scotsman of the night before, was on the way home.

"Are you on vacation?" he asked me.
"Not really," I answered. "Just tricycling to India."
'What—in that? What, no Cadillac?" The implied irony was no longer unusual when an Englishman met an American, especially on a scooter.

The costumes, especially of the women, now changed drastically. The Croatian peasant women were gaily dressed, some wearing pantaloon-ish skirts with red kerchief headdresses, looking darker and more attractive than the Slovenians. There were many gypsy encampments off in the field, with the usual colorful, large wagons and the adorned horses.

In Belgrade, once considered the Paris of the Balkans, I went to Hotel Moscow near Tito Street. It was Saturday night and Belgrade's main streets were crowded with idlers, most of them looking as if they were on their way to work rather than on an evening stroll. A bookshop I passed had a Yugoslavian translation of Virginia Woolf's *A Room of One's Own,* probably put in the window by a satirical bookshop proprietor. The signboards were plastered with posters announcing a Greta Garbo film, *Major Barbara,* Grace Kelly, and assorted American items of export much in demand here.

The shops had conglomerate things for sale, with impossible prices for all but millionnaires or commissars. The men's suits I saw were likely designed for scarecrows in the fields, not for proletarian backs and legs. A small calculating machine I looked at sold for almost one thousand dollars, though the price for it in New York was about two hundred. To window-shop was to suffer shock. For no matter how shoddy things were, the prices for them were fantastic.

The fragile-looking waitress who served me dinner hung around after she removed the plates. As I was about to leave, she said in a matter-of-fact tone, "Would you care to sleep with me for the whole night for 2,000 dinars?" It was not a whisper, but made like a speech; one short declaration of promises.

"Don't you make enough working as a waitress? Or is it love that you want? There must be a lot of love in Belgrade," I said, almost brutally.

"The government can't order love; and as for sleeping with

you, they won't object if you pay me in dollars. Are you an American?"

Now my country was being challenged; but my country, right or wrong, despite all of its alleged dollars, was not bait enough for me to accept this basic challenge to currency alone.

At breakfast I met the manager of a French singing quartet, the Frères Jacques. Over bad coffee we discussed the state of the French theater, American television, and especially boxing. The manager said that his company had come to Belgrade for one week and that he would be glad when it was over. When I asked him about Yugoslavia, he said, "I've only seen the highways, for I'm also the driver of the quartet. But I like New York. I spent a week there last year when the C.B.S. did a television show with us."

As I was leaving he yelled to me, "I also managed the American *Porgy and Bess* company when it was in France. That's when I learned about American boxing."

I left for Avala, ten miles from Belgrade, where the well known Yugoslav sculptor, Mestrovic, had created a Theban-like edifice with four heroic figures centered within the squarelike structure. A long series of stairs made up the approach, as well as the entrance and the exit. The park and edifice had become a popular picnic ground, for it surveyed the valleys and mountains for many miles around.

Later, I scootered down the steep mountain and had an *al fresco* Serbian lunch of goat cheese, spring onions, suckling pig, thick hunks of bread, and beer. Heading for Nis, I photographed the local landscape, mountains looming up in abrupt, irregular shapes. I felt heady, and almost intoxicated, with the smell of the blossoming wild locust; and when I passed some soldiers who were training German shepherd dogs, I found the contrast violent.

I had to go to Kragujevac before Nis, and the road, at least for my scooter, was one of the worst I had so far encountered. It was bad for tourists, catapulting with cars, and worse for a man bound for India with a frenetic if fragile scooter. I looped over mountain trails, observing signs in Cyrillic characters, which I had to memorize to assure myself that I had not reversed directions and was returning, via unrecognizable loops, to Belgrade. In the distance I saw Topolo, the mountain-top Byzantine church; but I was not heroic enough to make the steep, godly grade to view it closer. But though the road got worse, the people became handsomer and merrier, the shepherds and their womenfolk sat under the trees, with large black umbrellas by their side and portfolios which seemed to be part of every Yugoslavian's working equipment. The men were dressed in blue, jodhpur-like trousers, blue pullover sweaters, and blue split hats which looked like the summer issue for the United States Army. The women, quite sultry, wore white kerchiefs and sprightly colored aprons. The horses wore blue beaded decorations, their Sunday best, joining their human hosts under the shaded trees. In the small villages, however, everybody was out walking, since there was nothing else to do in these socialized, barren places. Walking became an exhibition time; for then the villagers showed off their clothes, their babies, their silence and their militarized mummery.

A boy passed playing a small violin, his movements a throwback to Chagall's ghetto scenes of Poland, but with local significance. The boy looked possessed, ignoring all the walkers as he played some wild tune that I could not make out. Its melody was broken, not continuous, as if he were composing it in a trance. In the next town, Cuprija, I expected to see other fey Chagall fantasies emerge from the fields of yellow iris; for nature had spared nothing of its lavish gifts here, though the totalitarianized men of Cuprija, with its mud streets and mud huts, had mud, not violins, for their private and public possessions. I wondered if the boy violinist had not, in some zany moment, worked out this musical stance as a protest against the mud-wallowing uniformity of his co-villagers.

Since reaching Nis was out of the question, for it was getting late, I headed for Aleksinac; but the village I stopped at was quite nameless and dark, lost in the Dakota-like, Serbian mountains. At a café a French-speaking priest playing chess interpreted for me, getting me a room, dinner and drink for under a dollar, sans comfort, and a game of chess, with the gentle, lovable peasants looking on as I made error upon error, and the priest, who was much closer to God, emerging as the purer player of the two.

"There is no mystery in chess; just in God," the priest assured me, offering me a vodka nightcap.

At Aleksinac, the next morning, I saw the police state in action, when my scooter was surrounded by masses of workers on their way to the grim factories. With the mist still thick, they paraded to work against a Walpurgisnacht backdrop. As the long lines of men came upon me having a 6:00 A.M. breakfast at a café, they stopped to examine the scooter, to ask me questions, to have a moment of chit-chat ... but an overfunctional if surly policeman yelled for them to go on to the factory; then he shouted to me to move the scooter back of the café, where it could not be seen. "It is keeping these men from going to work." With that he pushed into the milling lines, shouting like an animal trainer worried over his disobedient lions not going through the hoops in the lion cage. I hurried through breakfast, disgusted, yet saying nothing, for my three-day visa had long run out, and I was not looking for a lion cage or a Yugoslavian trainer to help me through at the border jump.

But the Yugoslavians play at all sorts of games. For in every factory town there is a modernistic, enclosed stadium; and Sunday, once given over to God by religious peasants and workers, is now given over to soccer, which everybody plays, including the little boy sheepherders, who kick at balls made of old clothes in their barefooted frolic along the trails and roads. They would leave their scooter-frightened sheep and goats, now clambering up the rock face of the mountains, to kick at their handmade soccer balls. Between kicking they would toot at their flutes like juvenile satyrs, inducing the sheep and the goats to return.

In the morning, breadlines formed in every Yugoslavian village and town, with shawled old women waiting in the rain for the large round, black loaf. But man does not live by bread alone. When they were not lining up for bread, they were searching for firewood and faggots, which, ironically, were usually piled up, fenced off, and guarded by soldiers much as if they were the staff of life. As for milk, I never found any, though I saw it carted off on donkeys, with the peasants riding sidesaddle, and the tins riding on the other side. But I found scallions, goat cheese, red peppers, tired aples, worn-out lettuce and a drink, which I remembered from my childhood, called kvass.

At Nis, I cashed a ten-dollar bill at the local bank, for it was impossible to find storekeepers with much loose change on hand. I had to be interviewed by the bank manager, who studied my passport, shook his head, remarking sadly, "Your visa's run out. What do you intend to do?"

"Cash this ten-dollar bill, if it's not too illegal. I'll worry about my visa at the border, or should I start now?"

"If it will make you feel any better, don't worry at all. I'm the worrying type, unfortunately. Is your bill not counterfeited?"

"Will it make any difference?"

"Who knows American money?" said the manager, counting out 4,000 dinars. His executive's salary, which was 18,000 dinars a month, amounted to only fifty dollars. He was shabbily dressed, puffing at butts of cigarettes, and he could have used Yugoslavian paper currency to roll his own, providing he had pouch tobacco. His clerks, unkempt and unshaven, were sour and cynical to the clients coming with small government checks, which they signed many times over before receiving their 8,000 dinars' monthly wage. The money itself was packed along the walls like old coupons, with total human indifference to the sagging, falling stacks, much as if it were phony money for the Yugoslavian political drama.

By catapult and somersault, I finally reached Skopje, a day's run from Greece, but five days late, which my passport showed only too indelibly. I had stretched my three-day visa into eight, which made me either a suspect criminal intent on ignoring Yugoslavian protocol for travelers, or a poet who had but a little lyrical scooter capable of just so many revolutions per minute. But Skopje got me into other difficulties when I scootered across a busy bridge that only allowed pedestrians. I was fined one hundred dinars, payable on the spot, to a traffic cop who said, "It is forbidden to drive any kind of motorized vehicle over the bridge during the day, though you are allowed to do so at night." Harassed by fear that I might be asked to show my passport, I paid and apologized, hurrying toward the Greek border, where I would have to face the inevitable question—"Why are you five days late? Your en-passage visa is clearly marked *three days only*." I fancied phoning the American ambassador to say that I was on my enforced way to some Yugoslavian slave camp, without my scooter; that he had better arouse the press of America and get to President Eisenhower.

But by the time I reached Gradsko, to lunch under a large tree, my abnormal fears evaporated. For I ate in the company of three young Turks bound home to Ismir. They were migratory workers or millers, though the way one of the German-speaking Turks said it, it sounded like *maler,* a painter. They drank my wine and I ate their spring onions; all of us happy to have a loaf of bread, a jug of wine, and spring onions.

A few minutes later two dark peasant boys and their tawny-haired little sister came along with their cows, and all had to be persuaded to be photographed. The little girl stood alongside of a baby calf, still unable to stand on its feet. Whenever the girl tried to stand the little calf on its feet, to get into the picture on four legs, her pantaloons came down and she would giggle. One of her brothers, with a show of modesty, scolded her gently, saying in Serbish, "You must learn to keep your pants up before a camera."

It was getting very hot, and I fashioned a havelock to protect my head and neck from the burning sun. The land was changing now, with many irrigation ditches to help cultivate the patchy, spare, bare land. Snakes wiggled across the road, always managing to get by before I would crush them with my scooter. Donkey carts with peasants wearing red havelocks clattered by, the carts loaded with oil drums used for carrying water. Another cart came by with rams horns, the symbols of another day. It was a parched land, with scraggy hills and desolate villages, as if the beginning and the end of geographic Yugoslavia hardly mattered; what counted was the lush middle, not the frontiers. I reached the frontier just as the Paris-Istanbul train thundered by and whistled, saluting me and Grecian freedom as it sped out of Yugoslavia.

But I did not speed out of Devdelija; not from the last fenced-off house of the Yugoslavian frontier guards and the customs. I was interrogated there by three men who were not content with my answers. I told the truth; my scooter was not jet propelled; the land of the Serbs, Croats and Slovenes was Paradise Enow, as scenery; my speedometer mileage showed 1,137 miles, which meant that I had gone over six hundred miles in their beautiful country; that even an American, speed demon and capitalist that he likely was, could not do better on Yugoslavian roads; that it would be a good idea to apply to Point Four for money to build up new roads; that mountains and valleys and hills and dales were mountains and valleys and hills and dales, obviously slowing up a man on a scooter.

They talked on. Couldn't I have changed my visa in Belgrade? But I hadn't, I remonstrated. I had eaten bortsch and *Kalbbrust,* much too tired to do anything but sleep. What about my photos? Had I photographed anything special, like airfields or depots? I hadn't even seen an airfield, let alone a depot. But I had, I confessed, held up ten trucks loaded with soldiers when I ran out of gas. Was that very bad? But the soldiers had laughed at me and my vehicle; so everything was quite all right, then.

Two hours later and over much tea, I was stamped out of Yugoslavia and ushered into Greece, exactly eleven days after leaving Genoa. I had seen my first socialized, communized, terrorized land . . . with much comedy, as I remembered Andro Spolar. Socialism had not raised man to a higher status. Good-by, beautiful Yugoslavia.

Chapter 3

THE GLORY THAT WAS

A MILE AWAY I was officially welcomed into Varday, Greece, by a nontotalitarian road, my scooter reacting with democratic *esprit de corps* in the presence of the thorough Greek customs. The welcome took an hour, for it included a survey of my finances, the serial numbers of my camera and typewriter, and their value; for, as the customs official said, "If these objects are stolen, then we can trace them for you."

"What? People bred on classical Greek literature steal?"

"No, no, no—it is only for your protection, sir."

I agreed that I was well protected from then on.

Over a glass of licorice-flavored *ouzo,* which he offered me, the customs officer asked, "Do you like sports? We have many famous Greek wrestlers, like Jim Londos in America. What is your sport?"

"Yugoslavian mountain climbing, which I've just given up. I'm going to be a Greek mountain climber instead," and I left, to assault the more classical Greek mountains.

Two hours later, much mountain fatigued, I quit for the night on the top of one. Nearby was an open-air café that looked more like a general store in Tennessee than a Greek bistro. Big barrels of *ouzo* and wine spaced the earthern walls. Ilya, the proprietor, unshaven, wearing torn pants, insisted that I stay for the night. So did his wife. He fed me, wined me, and announced to all who came in for a drink, "Meet my friend—he came all the way from America to see me." He introduced me to the young village doctor who spoke French; to a young farm hand who looked more like a fullback for Notre Dame; and to the local French schoolteacher, who was very reserved and literary, but with sweater-girl effects.

Our conversation took place in a mélange of languages, and concerned itself with England and Cyprus, a usual conversation piece, ending with the virtues of Greco-Roman wrestling as against Indian wrestling. Soon I was asked by the big farm boy to wrestle. When I said no, he grabbed me and we were at it, Greco-Roman style. We both took a fall each; then we became Indians, and it was even again. After that Ilya said, "You can't pull my finger away from my nose." I couldn't.

Anxious to get to sleep before they involved me in a Penthatlon, which appeared to be in the making, I decided to end all the athletic shenanigans by saying, jokingly, "If you can, collectively, lift me from the ground, I will make Ilya a present of my scooter. But if you can't, ladies and gentlemen, then I want the best bed in the house and the right to go to sleep immediately. Okay?"

It sounded so ridiculous that they agreed at once, with Ilya loudly planning a week end trip with "his" scooter. Ilya, the big farmer, and the girl who taught French conferred for a moment, then lunged at me. Ilya caught me by the waist, the farmer at the ankles, and the girl directed Operation Lift.

"Just one try," I insisted, knowing a trick that made me unliftable for about five seconds.

"Up, up!" yelled Ilya. "Can't you do better," he shouted to the farmer, who had almost split a gut trying to get me off the ground. Nor was Ilya doing better, straining at my waist to lift me bodily.

"He's rooted to the ground like a tree!" bellowed back the farmer. "Look, his toes have sprung roots." I could not be lifted up.

"Okay, gentlemen and lady, the games are over," I said,

breathing heavily from my own internal exertions. Years ago I had worked out exercises that slackened my body, adding many pounds to my weight at a given moment. It worked, fortunately, much to their astonishment and amusement. They searched my pockets, looked into my shoes, theorizing where I kept the lead weights they insisted were secreted on my person. When they pointed to my head, I said it was filled with Greek philosophy and Socratic laws, and that "nature abhors a vacuum."

Ilya's favorite phrase was *"Ich arbeite zu viel."* Then he would add, *"Arbeit! Arbeit!"* With that he pulled out a hoe, exclaiming, "I am in the field all day. So is my wife. And when my children grow up, they will be there too." He then offered me a bed in the room his children slept in, saying, "I'll see you at breakfast at five in the morning. *Arbeit! Arbeit! Arbeit!"*

Since they looked like Turks but acted like Greeks, I asked Ilya where he was born.

"Well, we're not exactly Turks. We were forced to leave Ismir, Turkey, in 1922. Everybody in this region is from Ismir, but we're Greeks."

I went to sleep on that, remembering that the Turks then bore few gifts to the Greeks.

In the morning, Ilya offered me his pistol as a gift, for he learned that I was unarmed and on my way to Turkey. "I know the Turks, Harry. You will definitely need a gun for protection."

Oddly, wherever I went I was told I would need a gun in the next country. The Italians said it was imperative for Yugoslavia; and the Greeks said it was essential for safety in Turkey, where, like the mule, it was a man's best friend. The Turks, in turn, said it would come in very handy in Iran, where the Persians were to tell me it would be a very special blessing in Afghanistan, along with the Koran.

I did not carry a gun. But I did carry a big stick to beat off the huge Greek dogs not bearing gifts, but baring large teeth; for on several occasions I was chased up and down the Greek mountains. Near Salonica a shepherd dog straying from its chores and flock almost became a hitchhiking canine when it leapt on my scooter to get a better bite. With my scooter wobbling dangerously, I managed to come to a dead stop up against a tree. I booted the beast, ran for a stick, flailing away until the sheepish dog went back to barking at his flock.

Hitting the tree did not help the scooter, which would not start afterward. A motorcyclist with a large sidecar, from which he was selling fish, produced a kit of tools, sold a few fish to passing lady peasants, and soon got me on the road again. Since he refused payment, I bought two of his high-octane-smelling mackerel, which I broiled in the open. The gasoline-flavored fish went up in blue spurting flame, cutting the cooking time by the explosive if unsavory spicing.

I was nearing Kastania Pass, which cuts through the 14,000-foot Pieria mountains with Olympian ease. It is a majestic mass, rugged and cloud enshrouded, with harrowing ascents and terrifying descents. I was hanging in the clouds, thinking of Olympian gods and of the poor gasoline mixture in my tank. The grades were straight up and straight down, much like a ski jump, and guard rails were missing at most of the hairpin turns. Complete sections of paving were gone, for the snows, rains and winds, on their seasonal rounds, gave back to nature what man tried vainly to cover over with his roads. Near me the cliffs went down thousands of feet to the great Macedonian plains.

I lowered the windscreen, which had become a dangerous sail at the windswept 14,000 feet. Suddenly the engine spluttered, then stopped; and I began to swear at the tampering fishmongering mechanic. But it was my own miscalculation. I had used up most of my gas going up in low gear. The next stop was at Kozani, thirty miles down the other side of the mountains.

I tried coasting, despite its attendant dangers, using the brakes for a few miles; but the brakes would not hold. I stopped, ate goat cheese and drank a glass of resin wine. How soon would a car come to bail me out with some siphoned gas?

I waited an hour, locked within these jagged mountains of the gods, listening to the winds disporting ancient Greek melodies, the past that was Greece engulfing me with antique associations. I saw dryads and nymphs, stone goddesses and fleshy ones emerging, imaginatively, from the unsculptured, savage stone. From this a civilization had been hewn; and no wonder, for Greece is almost all mountains; and to live, the Greeks had sculptured their way into civilization, cutting passes and places and greatness out of these mountains.

While photographing these memorable, marble mountains, I heard a familiar engine. A scooter was slowly coming up the pass. Straining over the edge of the cliff, I finally saw it round a bend, its two riders hugging on like Koala bears on a leafy gum tree. Dressed in khakis, they looked definably American, and when I yelled out, "Hi, fellow gods," I heard a return, "Hi, ya," in an Indiana accent, followed by a second "Hi!" in New Mexico-ese. They were two ex-G.I.'s studying architecture in Florence, scoot-touring for the summer through Greece's architectural past. They had notebooks, cameras, camping equipment, art books, and extra gas, which I happily siphoned into my tank; and the usual gasoline cocktail, which I swallowed in the sucking-in process. Now an expert, with my eyes dancing from the intake of the gaseous liquid, I asked the two expatriates to make me some powdered coffee, and the Grecian visions bubbling before my eyes vanished. Happy over this fortuitous meeting, we sang our way down the pass. At the bottom, the military police stopped us.

"Are you English?"

"No."

"Dutch?"

"No."

"German?"

"No."

"Then you must be Americans."

"Yes."

"Did you take any photographs up on the pass?"

My colleagues shook their heads. I lied, making a wry face, for I was still vaporizing with gas. It was considered a military area and I recalled the signs warning tourists not to take photographs. When I developed my innocent photographs a few weeks later, I discovered that I had some unusual pictures —two photos of marble slabs against a background of wild, jagged mountains, and a small rosebush against a brown rock, where I had drunk the powdered coffee.

At the Kozani gas station, which also served Expresso coffee, I asked the proprietor how he made Turkish coffee. Soon I became another sort of expert, for in his excitement he mixed the coffee he was about to hand me into the gas and oil mixture while filling up my tank. It took twenty minutes to clean the coffee out of the tank, during which time I drank five cups of Turkish coffee—but without a gas and oil mix.

But the proprietor knew his Turkish brew intimately, for he said, as we labored, "It's very simple to make Turkish coffee. You take two Turks, put them into a small cup, where they naturally grind each other to a fine powder. All we do is add sugar and serve. *Voilà!*"

"*Voilà!*"

When I left, I discovered that my horn, which had done yeoman service through the dangerous pass, was no longer sounding off its asslike braying. I bought a sheep's bell with a plaintive *F*-tone to keep speeding cars, straying sheep and Greek dogs from brutal encounter with my scooter.

Nearing Mount Olympus, which is a series of peaks abutting each other, I was stopped by two wraithlike children. Dressed in ancient togas, they dashed from behind a rosebush and minced about like dancers from Isadora Duncan's school of free bodily expression. Finally they bowed and presented me with a laurel and clematis wreath, crowning me the first Scooter King of Mount Olympus. When I started to leave these elfin guardians of the past, feeling quite regal in my headdress, I heard something strange, in accents commercial, and in the purest English, "This will cost you one dollar, please." I fled, but not before tossing them a few drachmas, somewhat shaken and bowed.

It took three days to scoot from Kozani to Athens over desolate sheep and goat country, past tawny-colored houses, thatched sheep pens and blue-domed Byzantine churches rooted in impregnable solidity. Near Petraha the mountains resembled primitive leviathans as they materialized fiercely straight up from the plains. A dead shepherd dog intensified the gaunt scene and made it less inhuman, with the belling sheep, softly sounding their chimes, indifferent to their recent guardian.

Near Servia I lunched on ten tiny lamb-burgers, my scooter and I surrounded by Greek soldiers busily talking like oracles from Delphi about England, Greece and Cyprus. When they stopped lambasting England, they played their version of baseball, and I was elected to referee the bizarre game—though they needed a referee in an armored tank. When a batter finally managed to hit the indoor ball, a fight broke out between the batter and the first baseman, who claimed he had been kicked where he shouldn't have been. As the latter-day gladiators took to their fists, I took to the road, fleeing up the quiet repose of the nearest mountain fastness, far from the tumult and Greek shouting.

The stony desolation grew, making endless brown arroyos, massive rock gardens and gullies. While slowly climbing a hill near Larisa, I met a large Gypsy caravan ready to camp for the night on the adjacent plateau, with their colorful tents and blankets emblazoning the stark, darkening sky. I joined the caravan, ate their shish-kebab, then learned the very energetic *horon,* a vigorous knee-twisting, waist-thrusting dance that I was later to witness along the Black Sea in Turkey. Exhausted, I crept into my sleeping bag. But it was not to be for sleeping, for all through the night the Gypsies played long-stringed instruments and blew on wooden pipes, rending the mountains with their imported Turkish lament.

Red-billed soaring cranes swooped over me as I approached the Narthakian hills, north of Thermopylae, where I rode for hours through rain, thunder and lightning. On two occasions I stampeded flocks of sheep en route to their enclosures, as well as a donkey with a sheep perched on the saddle. It was difficult to see in the constant downpour, but not difficult to skid, which I did when I braked suddenly to keep from running down the donkey. But what I failed to do, a bus unhappily did when swerving too quickly down the sharp mountain curve. The bus, loaded with peasants, furniture and livestock, ran into the donkey carrying the little sheep. Both were smashed, though the sad donkey had to be shot by a soldier from the nearby mountain garrison.

In the mountain fields the peasants took rocks and built little designs, as if to decorate nature, or to show how hard they must labor to make room for things to grow. But though nature is harsh, near here toga-clad gamboling gods once tossed gay thunderbolts across the snowy peaks, or so my schoolbooks said when I was a boy who believed in myths, gods and goddesses. But it took one long descending curve in time for the godly folk to disappear, giving way to Olympian sheepherders, who go up and down these arroyos and mountains to pasture their sheep and goats.

I was unable to find Mount Olympus on my linen map, or on the earth, for there are few signs in Greece for the self-appointed traveler. At Elasson, as I ate Greek salad under a poplar tree, I asked the inn proprietor, "Where is Mount Olympus?" He pointed to the gray clouds in the distance, and said, "Behind them."

"And the gods?" I asked, hoping to get a bona fide view from a man who had lived in the midst of the myths.

"Also behind the clouds."

Everything was behind everything; nature was behind itself, curving in and out, like the huge trucks coming up the passes, carrying loads of peasants and their furniture, with bright blankets shielding them from the cold winds blowing over the historical soil. The peasants moved from village to village, or from one mountain top to another, packing their bicycles and their kegs of food into the lumbering trucks, and off they went into contemporary Grecian poverty, looking for land to live and work on. They moved, unlike the tourists, waiting outside of cafés for the bus to start up again, staring at very fat Greek generals with two stars, which made them military men. The peasants moved amid the flashy American cars used as taxis, in a world that passed them by; for they were lost—lost in their current relationships to banks and to commerce, to all that made the world screech into their delicate, mountain-trained ears.

They seemed to prefer animals to humans; wild green places to civilized cities; to be forgotten than to be remembered—and here, in the mountains, one was easily forgotten.

As I neared a small village I saw my first sign in days. It said "India," advertising rubber tires, not the direction. An itinerant peddler passed me, carrying his wares on two donkeys. For these Greek roads are only good for donkeys; and though the Greeks built the Parthenon, they can't build roads through these mountains. The donkeys are the greatest natural architects of Greece today, or the most amazing travelers or hosts to travelers. They have a godly patience with their human burdens; a fixed idea of themselves; a carriage that defies analysis, and a strength that is truly Olympian. When I saw a small lamb sitting astride a mule, I knew that my feelings regarding the over-all values of the mule were only too right.

I was traveling slowly through these mountains, tired of the passes if not of the asses; of goat cheese and resin wine; of descending curves and ascending zigzags; of talking to Zeus, Daphne and Agamemnon; of Greek tragedy that faced me wherever I looked; of myths that were only too human if one wanted to enlarge on the contemporary scene, especially the fat military men who acted like gods but looked grubby and gilt-laden. But it was their world. It was not the world of the peasant walking with a baboon, followed by two Gypsy caravans; nor of the world of the wind over Thermopylae and the oleanders in the sun; nor of the sculptors who had hacked Greece into history; nor of Parnassus rising from the village of Brallos, wreathed in snow, without any sort of gods or humans, except a lonely ice cream vendor. And though all sorts of classical allusions came to me, I was soon the detached American lapping at the ice cream I bought from the wandering vendor. A minute later I was the historical conscience again, the many peaks of Mount Parnassus in veil-like shading making me tremble. The soft rolling valley below, with its misty hillocks, made the many-peaked Mount

Parnassus an incredible sight. The gods had left it to the poets, if not to the shepherd, to sustain its meaning.

At the foot of Mount Parnassus a double-track railroad sliced through, running to Athens. A stream of clear water raced down august Parnassus, to lose itself in shade and tall grass; and here goddesses douched between bouts of poetry and the flesh. Here, too, was a strange town of squat houses, with doors but no windows, with large General Electric signs plastering up the town's view and visions. Ah, poetry!

From all this, many centuries ago, a tragic sense of life had been given to man. Now, scootering over the classical land and in a more modern way of traveling, I was all too conscious how flesh ages and how marble remains, and how many ways there are of dying on a scooter. There was collision, with no defensive plating to protect you; a swerving vehicle hitting you while dodging a rock in the road; and the most awesome way, going over a cliff after hitting an unseen rock at night. I studied the forms of sudden death as I went through the mountains; for in various ways I would come close enough to know them well.

When I was a boy, my father used to make mead; and now, with classical and familial allusions, I had mead for breakfast, along with honey on bread, in Thebes. Pindar most probably wrote odes here, though Thebes has little left to live for again after its many tragedies and treacheries. Philip of Macedon had tasted military defeat from the Thebians and the Athenians; but Thebes had also tasted hemlock in its Boetian heyday. Alexander Pope in his *Dunciad* once wrote:

> From thy Boeotia, though her Power retires,
> Mourn not, my Swift, at aught our Realm requires.

When cities are gone, the loss is worse than human death; and Thebes sprawled in its latter-day artifacts, neon-lit, advertising on its stadium two kinds of soda pop that had replaced mead as a drink; there were "Nekki" and "Samson," both signs bulging with pictorial muscle men. In Athens I expected to see "Jim Londos" *ouzo* advertised on the walls of the Parthenon.

But the athletes and the philosophers were gone, and only the businessman remained, so stolid and solid, complete and exact, sufficient unto himself and the culture, occasionally getting sentimental over the ruins of Greece, but never like his forebears.

The policeman directing me toward the Athenian road summed it up, as police can when they lose civic pride. "This town is from a long time ago, and Thebes is not Thebes anymore. We have a museum, but it is not a very interesting one." Greece, so beautiful in the high sunlight, faded into a rocky vault at night. Thebes had been commerce in its prime, rivaling Athens when it came to keeping books, legends and religions. Here, too, Oedipus, the classical complex character who killed his father and married his mother, was born. The lyrics had run their course, and I scooted on, belling my way through the dense traffic, the mead, vinegar and honey mixed lightly in my expectations about Athens, a few hours away. Cadmus, the benefactor and founder of Thebes, had invented the Greek alphabet; but Cadmus had killed a sacred Ares serpent; and his family had known the anonymous vengeance and wrath of the gods; for Cadmus had killed something sacred. I killed a snake, running it down at Erithrai.

The approach to every modern city shows its smokestacks first; then the tenements; then the city itself, all finely unplanned. Modern Athens was not much different as I neared the blue Saronic Gulf, with wheat fields, small mountains, a stream, where I had a swim, then the belching smokestacks of Athens, at which I tossed the clematis wreath I had been "sold" by the two wraithlike children. There were no more Gypsies tenting under mountains, but Athenian traffic, gassing and gulling about Constitution Square, filling me with Greek fumes, with antiquity looming triumphantly from hills and slopes, streamlined endlessly for all camera-carrying visitors. Ionic and Doric stone columns, against which fleshier more Dionysian people leaned, gave way to time, to stone, to carvings, to eyes, to basic facts about architecture and beauty; people giving way to themselves in a tour of the moment, planned and executed, filmed, touched, breathed, known for a minute; giving way to the rendezvous between flesh and marble, to the eyes that finally glimpsed the clean majestic sky over Athens, where few birds flew; giving way to the ruined village, city, streets, hills; the broken, scattered Parthenon, soaring, standing, falling, complete and incomplete; the marble jettisoned and the marble standing, leaning into the winds on the hill; giving way to the senses marveling that all this still remained when the people themselves forgot to hold onto history or art or science or mathematics or architecture; that this is for the world, worldly and noble, graceful and brutal, terrifying with loveliness and cruelty; that stone is man's first and last touchstone until man ends in the earth . . . and if Greece is a museum, man will end there sculptured and alone.

This was Athens, without a real address; Athens rendered to myself, walked through for a week, and touched by everybody's vision and memory in a disconnection and namelessness; the sequences blurred but the tall columns standing there in the forever of the night.

I stayed at a hotel until recently called the New Angleterre. But with the British and the Greeks battling over Cyprus, the name disappeared one morning from the marble entrance, and it was now a nameless, faceless hotel, with nameless people stalking Constitution Square, looking at garish American magazines. While my scooter was being overhauled, I was being photographed for an Athens paper which ran a leader that I was going around the world on a literary safari, whatever that meant.

In Athens, however, it meant that one had to look up George Katzimbalis, Henry Miller's *Colossus of Maroussi,* the Edmund Wilson of Athens, but a giant of a man, and a man with gigantic eyes and all the heightened senses of observation and appreciation for life, art and letters, in that order. A friend from New York, the artist Kaldis, and the only man to wear a slight mustache on his nose, had written me a letter of introduction to George Katzimbalis; and at the American Bar on Constitution Square, we met over iced *ouzo,* though he was concerned with other *apéritifs.* Eyeing every woman entering or leaving the large square, Katzimbalis could nevertheless parry with zestful style, the good raconteur's sallies. He was wittier than most after-dinner speakers I had heard at charity banquets. But I remembered nothing except the amazing flow of language, none of which could stay with me, however much appreciated the verve and insight were at the time of utterance.

In 1955, Katzimbalis was invited to the United States, and he toured the country in hot August, traveling through twenty states and talking like all the Greek philosophers, though Katzimbalis specialized in Katzimbalis monologues. Though known as an editor of the Greek classics and as an idea-intellectual who stimulated others to write, Katzimbalis had done the unusual, or so I had heard: he had never written a word or put pen or typewriter to paper. He talked literature in fabulous swirls of language, inundating his listeners by the sheer flow of Katzimbalisms.

Inundated, hours later, I saw a familiar figure when up hobbled Kimon Friar, the New York teacher, critic, translator and poet, whom I had known years before. He sat down painfully, a cane across his stiff knees, sighed, then looked at us. A moment later he put out his hand and I remembered where we had met, in some vague recollection of friends past. The cane, jutting outward into the Square, was a symbol of a new beginning for Friar, who had turned toward God. It was life and death, I realized a few minutes later, when he told me terrible things about an accident he'd had a year before, while riding his two-wheel Lambretta scooter in Athens. It was the tragic city turned godly city, for Friar; but it also had comedy, as I learned while he sipped a lemonade.

"I was trying to avoid a donkey, when a truck, in joint communion with the donkey, hit me as I was riding on the outskirts of Athens. I've got steel pins in my right hip; and now, after eight months in a hospital, I've become a religious man," said Friar, squirming between God and the steel pins that held him together physically.

I had three hurried *apéritifs,* looking at the cane and the man, shuddering for my immediate future on the scooter. But

Friar displaced my private fears when he pulled out a publisher's contract from his portfolio, with Katzimbalis making jokes of a sort about a crazy man who refused to sign a wonderful contract.

"He's had it for weeks. Every day it's the same thing with him. Sign!" shouted Katzimbalis.

Friar was translating the new book by the famous Greek novelist and poet, Kazantzakis, author of *Zorba the Greek*. The new book was a tour de force, on Ulysses brought up to our times. When Friar finished his lemonade, he said, with an almost saintly expression, "The scooter accident put me on my back for eight months, so I was able to translate half of the thirty-three thousand lines. But my contract's terrible! The publishers want to offer me five thousand dollars as an advance, but they want fifteen percent of the movie rights, which is a hell of a lot. Imagine that!"

I imagined that, calling Friar a fool five thousand times over. "When did you last hear of a poet getting a five-thousand-dollar advance on anything? Shakespeare didn't make that all his life. You must have a criminal mind to reject the contract, Kimon." My last comment made all things very legal; for he shoved the contract under my nose, pointing out the small print in the process.

Dramatically, Katzimbalis borrowed Friar's pen and pushed it into Friar's ribs; and with some preliminary Greek chorus lines invented for the moment, Katzimbalis said, "Sir, you are a disgrace to poetry, if not to every inch of Mount Parnassus. Sign it or I shall denounce you for endangering Madison Avenue as well as Fourth Avenue, you killer of donkeys!"

"Then you think it's a good contract?" Friar asked meekly.

The comedy had gone too far, but he witnessed the signature, with Friar becoming shy, reacting with new reflexes that related the accident to the book on Ulysses; the scooter voyager no longer voyaging; the poet becoming the better translator; a pithy set of new perceptions arising from new circumstances. A new life and new conceptions, and all because a donkey and a truck had met Friar in moment of violence. Perhaps this was the nature of poetry—to collide? But the translator was the victor, not the donkey; and Friar had a new sensitivity.

Katzimbalis was going back to Greek themes, doing away with five recent civilizations, ten philosophers, fifteen famous ladies, famous for lying on their backs at great moments in history. Apparently they had a historical sense of timing, despite the absence of Mr. Toynbee to interpret the art of the ladies lying fallow. But Katzimbalis interpreted, returning to an old theme—Friar and his donkey: "Give Friar five thousand dollars and he'll buy up all Grecian donkeys for export to Turkey—and in Turkey, the donkeys can take revenge for the fall of the Byzantine Empire. That can be our Trojan horse."

My beloved donkey had suddenly become a menace and a weapon of war; but I did not dare defend the donkey, not against an expert of Katzimbalis' stature. It was easier to accept civilizations of a pre-scooter age, when travel was not international, and death by accident hardly existed. How many donkey riders had been killed during the early Olympiads? How many scooter riders were killed in the scooter age? Whatever the figure was, Friar was the symbol for me—somewhat crippled, yet a new man.

Another lemonade and *ouzo* later, and literature was safe again, the comedy done with, the contract mailed. As we parted for dinner, Friar asked me what was I doing in Athens and in what fashion had I arrived. I wondered how I could tell a man with a cane that I was traveling around the world on the thing that had crippled him and that I loved donkeys. But I told him, slowly, simply, factually.

His reflexes and voice said spontaneously, "Should I say anything to stop you or shall I pray that you come home alive? I'll pray." He did, right there, filling out the Square with his prayer, taking in my new solitude and my love for donkeys in the wine of the heart that his newly religious conscience dictated to him.

But I was not the only scooterer who needed praying for. When I turned up at the American Express I saw the two

American architecture students I had encountered on the pass. Both wore bandages and slings. They had run into a horse while rounding a mountain turn. I prayed for them, planning to be measured for a coat of mail and a hat to match, armored against horses, trucks and sudden stops, no longer the godly, geared Olympian. But I had an assignment from *Sports Illustrated* that nevertheless took me to Olympia to do a background piece for their Olympic Games issue. I went, sans scooter, which was getting its two-thousand-mile check-up—to sit in marble groves, cooled by the winds coming off the Gulf of Kiparissia and the Ionian Sea.

The forthcoming Olympic Games in Melbourne held many things for me; for I had lived there on three occasions. I knew it as I knew Paris: bohemian, semicolonial and conservative, built for man, with lovely women walking up Collins Street in the Australian's detached stride; a city that a man became attached to for its vigorous sort of living, if not its art. In any event, I was bound for Australia after India; for it was a red line on my map, along with Olympia, where the fun and games had first been held 1,700 years ago, when ancient man began to develop ethical systems of behavior in politics and sports. I went to Olympia.

Modern man's amateur standing as a gentleman Olympian athlete actually began in 1881, when Greek laborers and German archaeologists, via pick and shovel, disinterred ancient and forgotten Olympia, digging it up from 1,600 years of oblivion and twenty-five feet of mud, clay and sand. In 1884, Baron Pierre de Coubertin, a nonarchaeological educator and idealist, went the German scientists one better: Why not have the ancient games played again? The Olympiads had brought peace to warring Greece, at least during the period of the games, when every statesman, soldier and athlete was honor bound not to engage in military conquests; an interval when man shelved arms and took up philosophical and literary conquests.

With the flaming torch bringing symbolical, military peace, the games also brought cultural and religious festivals. Philosophers, poets, artists, sculptors and political sages attended the games, turning Olympia into a great center of art and learning. The athletes, too, were not just men with muscles, but unenslaved, free citizens without taint or moral tarnish. Actually, the victorious athletes were revered for more than their athletic prowess. When they returned home, the city walls were pulled down in welcome. For with such superior citizens, fortified cities were no longer needed.

Competitors came from distant Greek colonies, like Marseille and Sinope, arriving ninety days before the games opened. The spectators themselves had started for Olympia months before, pitching their tents outside the sacred walls of the sanctuary. With the spectators came the merchants, artisans and craftsmen, all tumbling about with their respective trades. The ancient cafés served goat cheese, mead and an early form of shishkebab. Flies buzzed playfully over the heads of the diners, who later offered up sacrifices to "Zeus, Averter of Flies."

It was like a country fair. Concessionaires sold wine, fruit, horses and statuary. Conjurers, acrobats and jugglers entertained the crowds by eating fire and swallowing swords. At the temple of Zeus, famous writers, sophists and orators plied their rhetoric. It was an amazing carnival FOR MEN ONLY; for women, who were not allowed to attend, had their own games at the feast of Hera. Menander, the comic dramatist, summed it up in five words: "Crowds, markets, acrobats, amusements, thieves," the latter hardly being included in the list of sporting events.

In the wasteland of Olympia's antique grounds, flowering among Queen Anne's lace, clover and yellow flowers, lie Doric and Ionic columns and pediments for tumbled gods. On less than a dozen acres are nineteen once holy, idolatrous places, beginning with the altar of Zeus and ending with the temple of Zeus. The seventeen other disinterred remains are the Leonideion, a guest house that once sheltered 340 important personages; the villa of Nero, still intact in a far corner, as if waiting for a musical visitor; the Prytameum, which housed the judges and arbitrators who decided upon photographic finishes; the Parliament, where the sages once sat; the Heraion, once a magnificent temple of Hera; the temple of the Sybil; the Exedra of Herodes Atticus, formerly a fountain decorated with statues of lusty nymphs; and the workshop of Phideas, who created much of the sculpture for the Acropolis of Athens and the playgrounds of Olympia.

But like all ancient places, Olympia is now a commercial enterprise; for the main industry of Olympia, as of all of Greece, is antiquity or replicas of antiquity, with six curio shops dominating the three streets that make up the modern village. The shops sell new Greek dolls and old Greek gods; painted vases stylized to represent eating platters of the Herculean days; handwoven wool bags and baskets, as well as canes for tired tourists wandering among the ruins. One enterprising shopkeeper gives away free *ouzo* as an inducement before offering up his goddesses in terra cotta. Even the gas station on Praxitiles Kondylis Street, named after the sculptor, is in the curio business. Flowers drip from a balcony above the gas tank and nine-inch statues of Hermes and the Winged Victory mingle with cans of Mobiloil, in a pursuit of the drachma that is at once sacred, profound and all-consuming.

Today, without thunderbolts from Zeus but to the echo of the braying ass, the modern Olympian dweller walks rather than runs; for he is far removed from marathon running, high jumping, the hundred-yard dash and all Olympic track and field events. And since he is not an athlete nor an aesthete, he is sometimes an agricultural worker in the many vineyards and citrus and olive groves. Or he works in one of the three factories processing olives; or, if he is lucky, he is one of the Tourist Policemen or semiprofessional guides to culture, herding his tourists over the Doric and Ionic columns of another world.

This is Olympia to its contemporary inhabitants. For the foreigner and the athlete, it is a sentimental shrine, a marble record of a civilization that once honored the body of man as much as it honored his intellectual life. Not oddly, the tomb of Baron de Coubertin stands at the entrance, near the Kladeus River that once buried Olympia under mud. And though the grandeur of an ancient marble civilization decays in time, in an age of chromium flamboyance the simple Doric and Ionic columns are all the more impressive. Man, real and fabled, built the Olympian legend and the shrine, the ideal philosophy between stages of military campaigns; an athletic-minded United Nations, but one that managed to make more than time-serving orators in the process.

Olympia was the real as well as the mythical Greece; it was ruins, carefully and lovingly sustained; and unless I went to the radiant islands in the Aegean, marbled monolithic Greece had been climbed by me mountain-wise; the ecstatic abstractions of nature's uncarved sculpture rivaling man's hammered-out art, seen, touched and known. I was outward bound, to lesser antique civilizations, but much more vigorous—with Turkey's lesser rocky road to careen over. And rather than retrace my way to Salonica and scooter over old ground, I took the S.S. "Abbrazia" from nearby Piraeus and sailed on the intoxic Aegean to Istanbul, only thirty sea-hours away.

I managed to convince four tough stevedores not to hoist my scooter aboard by sling which would have torn into the wiring and ripped away vulnerable parts; instead, it was hoisted on a wooden lift, raised much as if it were an overlarge, delicate clock. But at one stage in the raising, the man at the old-fashioned winch intentionally dropped the lift ten feet, coming to a stop before it hit the water. I jumped for the winch control and hoisted it myself, remembering how to handle it from my sailor days. The winch operator swore bloody murder at me, for I could have been caught in the wire coils and killed. But after the cussing-out and the laughter, in which I could not help but join, I was faintly praised by the four dock-wallopers for my odd talents, but muchly damned by the man at the winch, who was not a Greek philosopher nor a comic dramatist. In fact, he looked like a wrestler; but I didn't care to find out accurately.

I sat up during the thirty-hour voyage on the Aegean's memorable waters, the passing historical place-names constantly awakening some mythical association. Off to the starboard, lost in the glaring light, was the island of Lesbos. Below, miles to the south, were the Cyclades and wine-laden Naxos, once the

Dionysian well-springs for phallic orgies, natural living, Bacchic rites, choral songs and erotic dramas. Now it was just Naxos, another green island jutting into the Archipelago, its wine and its joys forgotten. Other islands had been the proving grounds for Greek tragedies; but modern war soon put an end to private heroism when the Dardanelles was bottled up by England. Modern warships, with their huge armaments, have displaced men of the heroic breed from the roots of Greek tragedy. Heroism came with wooden ships, not aluminum atomic devices or metal masses for massive destruction. The single individual hero was gone, leaving only myths about men who had lived in the blue-green world of Dionysius.

But there was Byron, who had swum, despite his club foot, with Lieutenant Ekenhead through a nineteenth-century Hellespont; his own peculiar stencil of the modern hero who had fought for Grecian independence, now too embarrassing for Greeks to remember. But all history is embarrassing years later, though it must eventually become myth or poetry if man is not to enfeeble himself with production charts. As I stared, transfixed by the phosphorescent hues of the Aegean, I saw myself going back over twenty years when on a similar day in June I had received a letter from exiled Leon Trotsky, then a historian of things Stalinistic, who was living on the island of Prinkipo, just off Istanbul. Trotsky could have been a great poet; instead he became a poor politician. He could have been a greater historian; instead he became a Fourth Internationalist, with all the bitterness of cliques and claques following him along the byways of Marxism revisited. My connection had been in the bitterness, in the Trotskyist clique, with an occasional letter from Trotsky regarding the nature of art and Marxism; for I knew how fraudulent was the Stalinist, communist, over-all Marxist theory that all art must be regimented and all artists literary stormtroopers, which Max Eastman and Leon Trotsky pungently campaigned against.

Trotsky, the exacting socio-aesthete gone Trotskyist, had lost all the battles, including the last one—when his basic followers attuned themselves to Stalinism in the hope they could once again be members of the Third International. But Trotsky, great writer that he was, had at least loosened up the language for many of us; and in the thirties I used to think that there were only three writers left for me: Melville, Louis Ferdinand Celine, and Trotsky; an American, a Frenchman gone strange, and Trotsky, the strangest of the three.

The usual gulls swooped and looped at the stern, diving for the remains of the fine Italian cuisine thrown into the sea after lunch. When the gulls left, a British carrier marched by and sent two Bats aloft. They swirled heavenwards, way above the birds screeching below the clouds, hiding from modern aluminum death, preferring the old Aegean but with modern Italian cooking.

At night, near Canak, an electric light portrait of Kemal Atatürk blazed on the Turkish shore under a yellow moon. Close by, the British cemetery to the "gallant dead of Gallipoli" made a strange, violent contrast. Not far off, soft blue minarets and pock-marked fortresses wound along the coastal scarps of the Dardanelles. The Turkish customs boat blew and came alongside, at which time the bar was padlocked and drinking ceased. I wondered about the mathematics of free enterprise, the three-mile limit, the padlock on spirits, and what agency in the United Nations world supervised such cultural barbarism in this directionless world.

But Europe was on the left and Asia was on the right, divided by equal symbols, politically and geographically; brown for Asia and green for Europe. Ah, hallucinating Hellespont! Xerxes had been the bridge builder, the amphibious general of the tight waters, building a bridge of boats to trample the elastic geography of invasion. Here Leander swam nightly to Hero, crossing over from Abydos to Sestos, two villages no longer shown on the map. Hero drowned, following her lover, Leander; and their drowning became an epic drama; a silent, modern, phosphorescent sea, which the local Armenians said "comes from red apple trees."

Hero was a lady suiciding for love, which ladies no longer

do; and Leander's Tower loomed up below the Golden Horn, jutting into the heroic romance like a Hollywood lighthouse. But no one on the ship knew at what point Byron had swum the four miles across the Hellespont; not the Italian captain, nor the Americans, Greeks, Turks, Persians, Afghans, English, French, Germans, or the Swiss passengers, all of them dancing, as they passed through the Hellespont, to American jazz, to the dollar swimming over all.

This was the entrance to Turkey. But where, actually, was the Hellespont? My excellent English map called it the Dardanelles, preferring piloting names to romantic confusion. But it was a body of water and some private drama, and a poet named Byron, a cripple, had done once for poetry, honor and fame what Leander had done many times for love. The hero was gone, the poets were on the couches of psychoanalysts— and the Age was TV.

What had Byron proven to himself, to his poetry or to his failures? Such failures! And I remembered, quixotically, Pushkin's jibes at Byron—but what a world, unmapped, uncharted, unpsychoanalyzed, they had swum through!

In *Don Juan,* Byron wrote this of his Hellespont adventure:

> A better swimmer you could scarce see ever,
> He could, perhaps, have pass'd the Hellespont,
> As once (a feat on which ourselves we prided)
> Leander, Mr. Ekenhead, and I did.

But Turkey had conquered the Englishman's sense of adventure. One of the passengers, an Englishman, told me that two of his countrymen had recently gone from Greece to Turkey with a scooter. Weary with scootering, they tried to sell the scooter in Istanbul. But though they had many prospects, it was impossible, legally, to sell it. After many attempts and days, the would-be-salesmen still had their scooter, which they refused, for reasons of sheer fatigue, to drive back to Greece. Exhausting their salesmanship, they hit upon an expeditious plan of ridding themselves of the scooter—to dump it off the Galata Bridge into the waters of the Golden Horn. This they did one bright morning much to the chagrin of the police, the Turkish customs and some would-be-customers. Not oddly, the Englishmen were arrested on the grounds that they had no right to dump it into the water. The Englishmen defended themselves with French logic. It was their scooter, the Turkish government would not allow them to sell it legally or illegally, or even give it away. And though it was composed of rigid physical properties, it could still have wafted away on a magic carpet. Also, insisted one of the Englishmen, had anybody actually seen them push it off the bridge? The arresting officer said that he had heard a noise, so it must have been the scooter falling into the water. The Englishman said it might well have been the sixth minaret that had disappeared just about that time from the Blue Mosque. With that the case was dismissed.

Chapter 4

SMOKESTACKS AND MINARETS

TURKEY HAD TURNED itself into a national lottery, with private and government banks selling chances on houses to the unhoused, never-sleeping Turkish population; for Turkey was trying to refloat its financial structure and construction through every variety of lottery imaginable. The streets of Istanbul were plastered with giveaways and gambles, or what a few liras spent on a lottery ticket could do, providing one won.

Istanbul was a bazaar; a subterranean world of Asiatic goods; an enormous flea market, but with more fleas. It was hot, dank, luminous, old and new, rickety and dazzling, abrupt and mystical, with the unfinished opera house off Taksim Square as a case in point; unfinished for almost fifteen years after it was started, with the singers losing their voices or getting too fat while waiting for the confusing scandals to simmer down and be transformed into singing operas, stages, audiences and applause.

But if there was no opera house, there was Abanoz Sokkagi, the red-light district ready for a musical score. On several seedy blocks, behind iron doors with large peepholes for extra-sensitive peepers, were more than one thousand girls and women, most of them nude or seminude, waiting for the peepers to decide and knock on the door. The public houses got their working girls from the plush streets, picked up as they solicited near the Hilton, Divan and Park hotels. Once in, the girl stayed there, imprisoned for a year, sharing her proceeds with the Madame and the city treasury.

It was a dismal, exhibitionistic district, with the peepers sampling before many doors, leering at the lewdness inside before finally knocking and entering.

The aged Madame, worn-out from counting money, since her contribution was rather vague, processed the would-be-client, her hands fingering an abacus, then a chit for the girl and one for the city—at the end of the night the three partners in sex tallied up, dividing each to each.

I had my camera and my usual assignment. When I finished, late at night—for I wanted photos of the street at its busy heights and lonely depths—I started for my hotel. On turning a corner, I turned too soon or too late. I was trapped. Two men grabbed me; one in front, with a knife to my stomach; the second, holding me pinned back, from the rear.

They started for my back pockets. But my wallet, with American Express checks and almost a hundred dollars in Turkish liras, was in my inside coat pocket. The man with the knife moved his left hand to snatch the wallet. I moved, too, kneeing him in the groin, then butting my head backward, to get the second man on his forehead. I was lucky, quick, accurate, and off, running up the street, expecting a gun to get me. Instead, I heard bloody murder, probably from the man I had kneed in the groin. I ran for a cop but there was no cop. I found a taxi, but the driver could not understand me. Disgusted, I made him take me to my hotel, where I told the waiter, who merely ambled along for a bottle of vodka and ignored my insistence that somebody go and get the police. Instead he sat down with me and laughed at my protests, remarking philosophically, "There are not many thieves in Istanbul, sir, so it is a privilege if they try to rob you. But they didn't, so why fuss?"

I fussed over six drinks—and went to sleep.

The next day I prayerfully salaamed at the Blue Mosque, at Saint Sophia, at the Obelisk of Theodosius, at the Galata

Tower, and at Taksim Square, where I had finally found a taxi, making five Moslem prayers for my intact body, soul, and wallet.

Istanbul is not a city; it is the middle of a musical comedy or a plot set to music that never ends. It wails from every street at night, from large open cafés bursting with feminine lament, as huge ladies, dark with singing sorrow, throw into the night passionate passages about love, mostly unrequited. For love is the only theme in Turkey, and it is love that is musically always tragic, despite the lyrics. One song, used by Eartha Kitt, goes like this in translation:

> When I was going to Uskudar
> I found a handkerchief.
> I filled it up with soft sugar.
> When I was looking for him
> I found him near me.
> I belong to him and he belongs to me.
> He is mine—who can say anything about this.
> I belong to him and he belongs to me.

But Istanbul belonged to the senses; with romantic flavors of love, touched by Western independence, making inroads into the villages. A village song called "Eninem," went like this:

> Under your eyes there are dark circles.
> I lead the sheep on the hill, and send my best regards
>   to Eminem.
> If your mother will not give you to me,
> I shall run away with you.

When a country starts building—and Turkey had built itself into the century with Kemal Atatürk's proddings—then it must suffer all sorts of social and economic derailing en route to looking like 1956 and not like 1556. American funds had been used up on great dams, on long-term construction rather than projects that would pay off in the immediate future. But Turkey, bustling and moving, was broke dollar-wise, waiting for Congress to send another one hundred million, which Congress was not doing. But the Turks were inventive; their dolmus, or share-taxis, were ready to take you across all of Turkey, if enough passengers were found to enter the doorless, fenderless, seemingly motorless relics that Istanbul uses for its transport system. The taxis ran on goodwill, not on gas or spare parts, which no one could get; for Turkey, dollarless, imported nothing to help its Eastern ways with a Western taxi.

Street signs were inventions of translators, who were likely unemployed poets translating between sessions of tea-drinking. One sign said, "Honest Toilet Soap." In restaurants they sold *Viski Soday, Votka, Bira,* along with shish-kebab, a usual dish all over Turkey, much like ham and eggs here. Chewing gum had replaced the polly-seed culture which even Suleiman the Magnificent had chewed on during his campaigns against the Greeks, when he established the Ottoman Empire. But Turkey, the former "sick man of Europe," was trapped by a lack of progress, living on an enforced American diet, yet very healthy despite America's dictation as to how Turkey should have spent its recent loans.

On the narrow walks running to Taksim Square, two men were carrying a large American refrigerator. In back were other carriers, humping huge packing cases that would floor a mule. The men had cords, belts, straps and a leather buttressing to take the strain off their back muscles; but they were beaten human mules who seldom reached forty before their legs, back and hearts gave way. Turkey was still in the middle ages with its human trucking.

An old friend, Hank, who taught at Robert College in Istanbul, took me to Prinkipo (now *Büyükada*), the "big island" in the Sea of Marmara. We hired donkeys and rode out to the beaches; we became Turks for the day, joining Young Turks dancing along the sands, with Hank, who had eaten only shish-kebab during his seven years' stay, saying, "In Turkey, everything is agglutinated—just like the language." After swimming, we were back on the mules looking for Trotsky's old home. It had disappeared, burned down, blown away, gone into history, with none of the local police recalling the Old Bolshevik exiled to Prinkipo. But I was still the sentimental man, freeing another segment of my past. Impulsively, I tasted the sea and the sand of his exiled place.

Hank, who was not an adventurer, pointed to the Istanbul Hilton hotel as we were returning to Istanbul. "You know, I've never been inside."

I changed my residence after trying two cheap hotels, their native fleas, their imported bedbugs, their portable furniture, their unexportable heat—and off to the plush Istanbul Hilton I went. A street guide who had followed me around, unhired, changed my dollars at a four-hundred-percent-higher rate. I got twelve liras for the dollar, though the official rate was under three liras; and the Hilton soon housed me, dirty khakis and all.

I salaamed as I entered, with two porters, who looked as if they had just come out of the sixteenth-century, lugging my two canvas bags and camping gear. The manager fretted but I got a room overlooking the Bosporus at two dollars a day; and Hank and his wife, Evelyn, finally saw the insides of the Hilton: the manner of the American imagination gone Turkish and Arabic; a commingling of architectural tastes, functionalism and flair, embracing this living American minaret on the Bosporus. From personnel to furnishings, it was a vision of Turkish agglutination; and I expected to see Suleiman come out of the manager's office, or Mohammed II emerging from the coffee lounge, where pretty pantalooned girls in twelfth-century garments served demitasse, as they danced-walked with their burnished trays into June, 1956.

Two days later, I drove the scooter up to the Hilton and completed the contrast. There I was, my scooter parked at the door, a doorman caught between bowing me out or throwing me out. I packed my canvas bags and started for Asia, just a ferry boat ride across the Bosporus. I left without flutes, drums or dancing girls to escort me away; but a wheezing motor, which kicked over then stopped, ruined my exit from the Hilton with some gaspings and some bangs before it died on me completely.

I was soon surrounded by experts all talking Turkish. The *yag* and *benzin* was wrong, said one; I should have mixed more oil with my gas. Perhaps there was not enough air in the tires, and ten people stooped to look, feeling compelled to complete this pneumatic offering to aid a stranded American. Then, perhaps it was the motor, which could be taken apart by a brother, said another who had a brother who was a mechanic, but at the moment in Ankara. Could I wait a few days? I couldn't. After that it was the spark plug, the brakes, the flywheel, the magneto, the heavy air, all said in Turkish, kindly, smilingly, softly, but not helpfully. Finally a man who said that he had studied the humanities at the University of Wisconsin came along; and we talked about William Ellery Leonard, who had been his teacher; and we had friends in common on the faculty. I had, at last, found a man who was not a mechanic, though he knew of one.

He phoned and soon a young mechanic came. It was solved in a moment—the emergency brake had been on all the time. I was the red-faced American who had tried to make a sixteenth-century exit from the Hilton, but had forgotten to release the emergency brake.

The mechanic, Cahid Alakan, now turned out to be an amateur boxer, and he immediately made me his manager. "Could we postpone it for a while?" I asked, looking at Asia burning brightly across the Bosporus. No, we must sign papers now, insisted muscular Alakan. With the Wisconsin graduate gone, I pulled out my notebook and wrote two copies of the following: "I, Harry Roskolenko, currently in my right mind, hereby become the manager of middleweight Cahid Alakan, and promise to get him into Madison Square Garden within a year. Should this prove to be somewhat impossible, then into Yankee Stadium. Mr. Cahid Alakan insists that he wants no money—he just wants to fight in big places. In any event, he is a very good mechanic, for he knows when an emergency brake is on, which I don't know. Harry Roskolenko, June 19th, Istanbul, Turkey."

I finally arrived in Uskudar, Asia, the manager of a stalwart middleweight willing to fight in the heavies, if it came to a pinch. To my right was Leander's Tower, standing sentinel to the legend; to my left streamed a long line of birds, locally called "the wandering souls," heading toward the Black Sea.

In back, swimming in the noonday sun, were Saint Sophia, the Blue Mosque, the Seraglio, the Mosque of Suleiman, the Galata Tower—Istanbul new and Istanbul old, self-contained within its European fortress and fancies. But I was on the opposite shore, ready to scoot into Asia at last.

It was wilder, rougher country, with minarets and smokestacks joining cypress trees, edging out the past to contain the brief beginnings of the century. Old huts of crisscross wooden slats sagged alongside new pink houses in conjunctures of time and progress. But the minaret was Asia, the smokestack America and Europe, the cobblestones more of Europe, along with the occasional stretches of fine macadam. At Ismit a long veiling mist hung over the bay, creating imagined blue and white mosques in the sky. Actually, it was smoke from a factory that the wind had blown up like white tasselings across the sea.

But it was Asia and Asians, the Turkish women still pantalooned, wearing bright havelocks, many-skirted as in the harem days, stooping in the fields of wheat, while men idled or supervised. Soldiers, badly dressed, rode by in horse-drawn wagons, much as if this were still the days of the sultanate, before the jeep came from America.

Two large bulls locked horns, making a terrifying noise trying to break apart or slaughter each other; yet the violence was hidden in man, so kind in his spoken ways, in his rich projections of goodwill. It was violence engendered and remembered of the Terrible Turks who had streamed into the Middle Ages from some tempestuous Tartar beginnings. Now they were walking quietly along the broken-up roads, drinking *tay* (tea), sitting under cypress trees, saluting a stranger in a stranger vehicle. It was the violence of a compressed history, for the Turks had arrived late on the scene, having been caught between a Pagan birth and a Moslem re-birth on the decaying soil of the Byzantine Empire.

It was like the night I had first entered Greece, at Ilya's pandemonium inn, wrestling and drinking; a violence gone to seed, but an anger remembered—and this was Turkish Asia. Ahead of me was a bus caught in the middle of a big herd of cattle trying to cross the road. I detoured, amid the trumpeting and the bellowing, for it would have looked even stranger to be chased by a bull. The violence of the bull sounded over the fields, between the soft brown hills, breaking into the old silences by its normal intrusions.

Two hours later I had a belated lunch at a café, waited on by five child waiters, each bringing me one dish from the five different groceries, running after shish-kebab, cucumbers, bread, tea and wine, in an elfin enterprise; a restaurant that had nothing but chairs, tables and cutlery for sudden diners. But Turkish cafés are like the exaggerated landscapes, either bare or overadorned with Atatürk symbols. They usually have a grill, tea glasses, a picture of Atatürk, some large tomatoes arranged like a Longchamps window, and bepatched clients. The center of the cafés are the tea glasses, and portraits of Atatürk; all else being superfluous, including the food.

Near Kizilcahamam, I met a Turkoman caravan, among the last nomads of Turkey; their bullock-drawn uncovered wagons, built like wicker baskets, contained all they owned. The wagons were laden with brown drinking gourds, copper pots, striped bedding, firewood, with chickens and birds in yellow baskets riding on the driver's seat. The women covered their faces when they saw me, but the men and the children waved quietly. The young, unharnessed bullocks trotted along poking their heads into the harnessed bullocks. Dogs, tethered to axles, ran freely under the wagons despite the ropes.

It was a caravan going nowhere, though Turkomans have destinations—grass and water, the public lands to graze their flocks; the flocks, like their masters, drifting in an endless motion, unencumbered by their masters' lack of land deeds and titles. It was just wagons and animals, women and children, finding land and water on a free ride to eternity. But once the government subjected the Turkoman, the world would be a loser, too.

As I saluted back, the children smiled happily, as if I were a big American brother also drifting through Turkey's engaging landscape and history. Xenophon had been through here; and though all others had been conquered and reconquered, the

Turkomans went their ox-drawn way, not crowding into destiny or commerce.

I spent the night camping a hundred yards from their encampment. Some pitched tents; others lay on the ground on bright blankets. Little fires flared up for cooking; children's voices rose above their elders in little sound games. But mostly it was talk done with; acceptance had been proven, men and women knowing their roles, the children still learning theirs, the animals grouped and snorting, the fires sparkling—and the music from the flutes creating an innocent lament, not complaining, not pained, but merely setting off an echo of the Moslem world they also inhabited.

A few of the older men came over. I was offered tea and *rakkı* (a potent, anise-flavored drink); my hand was shaken many times as we tried to communicate, making words that always ended with a laugh. I tried out my fifty words related to bread and food, to water and the amenities or questions about cheap hotels, the roads, or where Mount Ararat was—a place which haunted me. One of the Turkomans said he had once passed through Dogubayazit or Mount Ararat, which was close to the Soviet Union and Iran; and though the land was bare, there were many Turkomans who had settled on the land around Biblical Dogubayazit.

After Kizilcahamam, the villages became more primitive, with clusters of one-room, windowless shacks dotting the hillsides. The women slapped their washing on stones at the nearby streams; the rice fields gleamed. The snow peaks of the Köroglu mountains cooled me, though I was not too far from the small Köroglu desert, and the heat trapped me into dozing. My speedometer reading had passed the 1,800-mile mark the day before, and I felt mystically related to the machine, the relationship including the callouses on my palms. I was going where I wanted to go, and the deserts were part of the journey into myself and out of myself. To my right were fantastic rock formations, much like Dakota's Badlands; but there was nothing before Ankara, the capital of Turkey—just the stark land and black toucans with long yellow bills sitting on telephone wires.

At last—Ankara, the modern, marbled city that Kemal

Atatürk had created between the brown burning mountains. I stopped at the Cihan Palace, being taken there by a young enterprising Turk named Mehmet Ali, who proudly showed me the American Bar, the Roman toilets on my floor, the telephone in my room, the windows that opened on the busy street, the buzzer that called the waiter, who never came, and the lights that worked on push-buttons that I pushed but which did not put on the lights. Yet Mehmet Ali was proud, instructive, loquacious and in the know, all of which I discovered during dinner at a place we taxied to, a garden restaurant he endorsed because, "It is very good because all the members of the government come here, with pretty girls." I saw the pretty girls, the alleged members of the cabinet, and we ate shish-kebab.

Mehmet was more a conspiracy than just a man; but then, Ankara was a natural place to be a conspirator, for Von Papen had set the norm for the cloak and dagger profession. But Mehmet was more than that; he was also a self-styled public relations man seven thousand miles from Madison Avenue; for he had a built-in receptivity and sold Turkey to me at the rate of one hundred American-learned words (Robert-College-learned) a minute, with a slight touch of Baclava in his sales recipes. He came from Kayseri, which, according to public-relations-man Mehmet, was the fastest, mostest, greatest, grandest, beautifulest, industrialistest city in Turkey.

Four minutes later, Mehmet said he was getting married in a week; that his future wife came from the richest family in Kayseri; and could I, please, be his guest. Yet he was not bragging; he was merely being an Oriental. Some exhibited their status in clothes and jewelry. Mehmet talked of it, fashioning verbal diamonds toward the same image—richness. But rich or poor, I unfortunately had no time for weddings.

Animated, excitable, active, Ali and his marriage would be an asset to Turkish industrialism, with many offsprings assured, though Ali was not an industrialist. He merely owned, he told me, three antique shops, three cars, and three farms, but his future wife's father was an industrialist who owned almost all of Kayseri. Since I owned nothing but the scooter, I suggested that it would be good for my old age to own something tangible, like land and a house. Mehmet promised to get me both if I went to Kayseri as his wedding guest, and I wouldn't have to drive there on my scooter. He would put it on a truck and I would ride in one of his three cars.

Mehmet Ali was more than merely enterprising. He was a veritable self-appointed cultural guide for all my known and unknown needs. He called on me early in the morning to shepherd me around, but I begged off until the afternoon. I had some breathless chores, like finding a tube of brushless shaving cream; and there was only one place, the American Army PX near the American Embassy, the cluster of buildings that made a little America within sprawling Ankara.

I spoke to a sales girl at the PX, who ushered me into the manager's office. But the manager was elsewhere. I repeated my request to his secretary, showing her the new cut near my left ear, received that morning by using Turkish, unlatherable shaving cream. I had a tough beard, and she felt it, saying, "You have very good color. What are you doing in Ankara?" I mumbled something about going to India, then around the world, hoping that would soften my request for shaving cream. "A man needs what a man needs; and American habits are hard to break away from, especially shaving habits and styles." And, I said, "The poor Turks have only shaved since Kemal Atatürk; and brushless cream is not obtainable until the new American loan comes through. Shall I wire my Congressman, or shall I grow a beard?"

Eventually an assistant manager was found, who went to see the commanding general on the second floor. The man with one star took forty-five minutes to make a decision. When the assistant manager returned, he shook his head and said, "I am sorry, but you cannot buy brushless shaving cream at the PX unless you are a member of the Armed Forces. So says the General."

"Tell him to go to hell! So say I." And I stalked out, running into a second lieutenant who bought me a large green box of Mennen's Brushless Shaving Cream as a gift. Glory be to Mennen and to dappled lieutenants.

At four, Mehmet came to the hotel to take me to Atatürk's tomb. In the car was a young Turkish lieutenant, Nuh Kusgulu, an economist trained at the London School of Economics but working for NATO in Ismir. Lt. Kusgulu complained that during his five years in London he hardly made an English friend. Was there something so unusual about a Turk that kept Englishmen from learning about Turks? I said that the best modern book I had read about Turkey was written by an Englishman, M. Philips Price, a former correspondent and civil servant. Lieutenant Kusgulu, hardly pleased, scoffed, "They write good books, but I lived there for five years without getting more than a pleasant greeting, even at the pubs. What a people!"

When he saw the tomb of Atatürk, an enormous structure of marble and stone, with scarcely any visitors, he grew very angry, especially after listening to Mehmet, who was now overselling the tomb. Lieutenant Kusgulu sneered, then shouted in boiling rage, "If Atatürk was alive, he would destroy this hideous thing. Do you realize that eighty-four million liras were spent on this? No wonder the Americans are suspicious!"

Mehmet was not impressed by Lieutenant Kusgulu's economics. Mehmet thought, loquaciously, that the green marble was beautiful, the brown stone more so, the gold mosaic very golden, and all of it quite cheap for the price.

Lieutenant Kusgulu turned to me and asked, "What sort of a tomb does your great Lincoln have?"

To say that it was a simple tomb was to take sides. So I said, cagily, "Heroes and statesmen are not architects; they never see the edifices that are named after them," whatever that meant. Mehmet sneered this time, for Lieutenant Kusgulu felt I had taken him off the hook.

From Atatürk's tomb, and at the insistence of Lieutenant Kusgulu, who wanted Turkish reality and not its marble fictions, we went to the red-light compound, where 128 women averaged 600 lira a month cohabiting on three quaint, hilly streets. When not at work, they stood outdoors and cooked over braziers or made clothes or fought with each other, their voices singing into the crowded streets as if they were selling halavah, not themselves, in this competitive, self-employed enterprise. When a man approached a group of four prostitutes busily cooking over a spit, and insisted that one of them take him to her bedroom, she asked, "Can't I even finish my dinner?"

"If you do, you lose me as a customer. My wife is waiting for me."

"Oh, well, I can't keep your good wife waiting. Let's go."

We left, with Lieutenant Kusgulu remarking, "I am disgusted with Turkey and I am migrating to Canada when my national service is up. I intend to marry an English girl and together we will go where there is some scope for us. I am a Western-minded man, not a Middle Eastern antique," and he glared at Mehmet, who merely laughed at the anger of his good friend.

"You are both having dinner with me," said Ali, later, driving us to his large, walled-in house, where we sat in the garden and met his sisters and brothers. It was a Turkish family, tightly knit, with a sense of belonging. They had three old servants who sat on a large bed near the garden and stared at me, much as if I had just arrived from heaven on a magic carpet. It was always my eyes. One got up to look closer, then whispered to Ali.

"What is she saying, Ali?"

"She wants to know if you paint your eyes blue. What shall I tell her? She wants to do the same."

I laughed self-consciously. "Tell her that my mother gave them to me," I said, thinking this to be sagelike.

Ali interpreted for the old servant, with Ali shaking his head, hopelessly confused by her next request.

"She would like to meet your mother, she insists. Can that be arranged?"

"Very soon, I'm sure—in heaven."

One of Ali's younger brothers, Metim Kuran, had just returned from a two-month cycling trip around Turkey. He had lost fourteen pounds, had lived with fishermen, peasants, herdsmen—and the whole trip had cost only 123 liras, or about twelve dollars on the black market. The dinner was for Metin Kuran and Harry Roskolenko, fellow adventurers. Mehmet, as the oldest brother, made a pre-dinner speech, selling us Turkey again, with an added dash of vodka to start the dinner now being put on the huge table by the old servant, who still stared at my eyes, shaking her head in disbelief every time she looked at me.

We were stuffed with stuffed peppers, stuffed tomatoes, stuffed squash and stuffed *baklava* (a dessert). The moment we stopped eating, Mehmet became the frantic organizer, directing all his brothers and sisters to put on their street clothes because we were going to Ankara's large Luna Park to see an American art exhibition at the permanent American pavilion.

There was no choice; Mehmet wanted art. There, at the streamlined, busy pavilion was American art—engravings and woodcuts, including a lithograph by an old friend, Frederico Castellon, whom I had met at Yaddo in 1940, and who later did an illustration for one of my books of poetry.

"Who is this one?" snapped Ali, pointing to a piece by Eugene Berman.

"What does it mean?" asked Metin Kuran.

"How do you look at it?" asked one of Ali's pretty, buxom sisters. "Do I stand on my head?"

"It is very pretty no matter how you look at it," said Ali. "It has dazzling colors, yes?"

One of the drawings, a nude, was even more buxom than Ali's sister, who stood before it examining every part of the nude's anatomy, then looking slyly at herself to see if all things equaled out by comparison. Ali pushed her away, uttering some rebuke, and we all moved on, suddenly lost in the veiled Moslem attitudes, especially when they concerned the naked female body.

After two days of Ankara's semi-oriental graces, I left for far-off Erzurum early in the afternoon, riding through dusty, brown, slaty, adobe, donkey country. It was a garden of rocks and thistles, and I wondered how people scratched a living from this painted-desert-like country with enormous garnet-coolred boulers that looked like idols squatting on the parched, ancient earth.

Everything was saved by the peasant. Drinking water was paid for by the bottle, as I discovered when I lunched at Elmadag and the water-boy waiter added twenty kurus to my bill. Whenever I refueled and a few drops leaked from my reserve gas can, there was always a man with a cigarette lighter bending down to catch the drops.

The road followed the muddy Kizilirmak River, which looked like streams meandering to the Mississippi through well-padded America. But here the analogy ends, for Turkey and Greece are peculiar surface civilizations; they have their formidable pasts, but now they live off imports, even importing roads, airfields, fuel, and all the motorized vehicles that scratched at their earth. Remove a few roads and vehicles and these countries go back to the age of the ass, back to their mountains and stones.

The Turks in the country were more barefoot than shod, more obedient than believing, yet very kind, confused, and good. Somehow the land had created these stoical people, and their governments exploited them and their stoicism as well. The Turks wander about whiskered and patched, much as if their second-hand clothes, which are really a series of patches, were hand-me-downs for a beggar's opera. The man and his clothes are immensely charming, for he wears his poverty with zeal, grace, and dignity, though why should one be gracious about poverty? I soon discovered that the second-hand clothes import business operates like a cartel in Turkey; for since few Turks can buy new clothes, old clothes become a major item of import.

Around Delice the land began to change. It became cubistic, with square fields of various colors and square squat houses breaking into a many-colored vista. The mountains were round and fat like dancing girls. As I passed a mosque I saw a stork alight on the roof where its nest was. Other birds flew over this coffee-brown earth, but their green-tipped wings were too fast for my camera. The whole area, with miles and miles of newly turned earth, reminded me of Carl Sandburg's lines about the "slabs of the sunburnt west." It was soft, rich land; the peasant women in red costumes that matched the earth seemed happier and less burden-laden than the women I had seen in other Turkish villages. Enormous seas of grain stippled the rises of red volcanoed hills.

But what is a Turk like in his village? Since I often watched them sleeping, talking, dancing, praying, and even working, I compared them to others. I realized that each country has its special, national way of waking up. In an English village it is casual, understated—with tea and talk about flowers. In France it is philosophical, sober, but with Gallic sophistication. In the United States there is but one thought and one statement: "Where is my cup of coffee?"

But in a Turkish village it is a harum-scarum hustle of men, mules, chickens, goats and sheep, with little boys out-shouting Turkey's livestock. It is cumulative and dissonant, as if an opera of men and animals were being composed at 4:00 A.M. In the village of Sungurlu, where the mule is man's best friend, life begins at 4:00 A.M. For Turks apparently never seem to sleep, even though they appear to be in a resting position twenty-four hours a day.

It is morning. Howling peddlers and assorted beasts of burden are congregated in the square opposite my hotel. Little tea-boys with brass trays are busily delivering petite glasses of strong Turkish tea. Their customers are everywhere about the square. Some are waiting for a bus to take them to Ankara, and their baggage, a makeshift of bundles and ancient valises, is piled alongside the equally ancient tables and chairs of the café. Others, belonging to Sungurlu, are merely inquisitive early risers who have already made their prayers and are now assessing the would-be-bus-traveler's clothes, baggage, personal effects and personality. For every day is a holiday. The native is doing no more than he usually does. He will have his tea on the hour, say a few words to his equally sentimental neighbor, and, between tea talk, listen to the whining, undulating, lamenting, sinuous music blasting from a local radio station.

The hour of 4:00 A.M. or 5:00 A.M. has no meaning for young Turks who like to dance, especially a dance like the *horon*. For our about-to-leave bus will have a group of four dancers who won a local competition the night before; and now, while waiting for the driver to finish a sixth cup of tea, they decide to dance. They do the *horon*. The four young men join hands, and quietly, as if restraining an emerging violence, they begin. It is an intricate dance with Russian-Greek folk choreography, though it is obviously Turkish. In a straight line, their knees somewhat sagging, the dancers go from one subtle pattern of sudden bodily shifts, bends, stances, gyrations, and oblique dashing movements to frontal, quiet interplays of rhythmic motion. They tap, they swerve, they arch, they bend, interspersing swift Russian *kazatzka* steps with slow sinuous turns. Neither the other passengers nor the tea drinkers are surprised. Why shouldn't people spontaneously dance at 5:00 A.M.?

The bus is ready to leave. Forty people pile out of cafés and hotels, hand their baggage up to the driver, who is perched precariously on top of the bus, and dash off for a seat. Should eighty travelers make for the same bus, eighty travelers will somehow find room, if not inside, then on the top along with the swaying baggage and occasional basketed chickens, goats, or sheep. For Turkey travels like this; it is an adventure, a mission, a study in how men, women, children and animals go from place to place.

The bus leaves, and with the "foreigners" gone, Sungurlu settles down to its normal occpations and preoccupations. Little barefoot boys, much bepatched, with beaming eyes, try to outdo each other in selling their quaint Turkish pretzels. Others will come by with bottled soda pop called *gazus*, or with caramels, packages of cigarettes, matches; then the inevitable shoe-shine boy with his elaborate cone-shaped brass box that contains all of the shoe-shining equipment. He like the others, will plague every man within reach until the men buy something or have their much overshined shoes shined once again.

Now tarbushed peasants and their shawled women shuffle into the village leading a few sheep and donkeys. The men are a study in humility, patches, and mustaches. When a sheep gets out of line, the peasant whacks it with a stick. When the donkey refuses to move, the whack is a bit harder; for the donkey, almost invisible with a load of hay piled from ear to tail, can hardly see his master and is only reminded that he is a donkey by the whack of the peasant's stick. In the middle of this, the muezzin will sound his awesomely sad call to prayer; and the peasant will leave his mules, take off his shoes at a fountain near the mosque, wash his feet, and enter the rug-covered interior. There he will sit on his knees, and, his head and body swaying toward the East, will touch the floor, with his forehead to the carpet as he chants in prayer, his voice blending with that of other peasants. Later, having made the first of his five daily and nightly visits to the mosque, he will rejoin his donkey and sheep, take them to some public grazing ground, and sit under a tree, smoking his hubbly-bubbly pipe. Or, if he is very tired, he may go to sleep, waking up in an hour or so to find that his animals have wandered off.

Let's look in on Sungurlu just as the stores are opening and some shoppers are making an appearance. A woman in very gay colored apron, red and purple under-skirts, shortish black pantaloons, and gold jewelry, is buying a liter of yogurt. Another woman walks by, carrying on her back a cradle. In it is a one-year-old child; for the woman, having no sitter, is taking her child to the fields where she will hang the cradle on a tree and begin her fourteen-hour day of field work.

At a café, a patron is invited into the kitchen to select his breakfast. He will settle for white goat cheese—called *beyaz peynir*—ripe olives, *ekmek,* (bread), and *bal,* a golden honey. With that he will order either *kahve,* which is Turkish for coffee, or *tay,* which is tea. He will get some lumps of *seker,* or sugar, to sweeten his brew. And as he eats, the dolorous, whining radio will get louder and louder until all of Sungurlu will literally echo with the nostalgic, eternally lamenting female voices broadcasting at all hours of the day and night.

The dust is rising. Now the proprietor or one of his many boys will sprinkle the café floors inside and outside with left-over tea and glasses of water. This is a ritual performed with every left-over glass. Should some tea or water sprinkle a café-sitter, he will think nothing of it. He will merely continue rolling the yellow beads which he always carries with him and pulls out when he is bored. Some beads have a religious meaning like a rosary; others he wears just to finger by way of diversion.

In another café some men already are playing checkers, although it is only 6:00 A.M. For Sungurlu, like every Turkish village, has its quota of checker cafés, where intent-eyed men, who seem to have lost their occupations, click away their waking hours.

Should a stranger, an American, suddenly drive into Sungurlu, his car immediately will be surrounded. Even though gainfully employed, men and women, boys and girls will leave their labors and hustle over to the American's car saying "Hello" and "Okay." They will admire the car, whether it is a jeep or a Cadillac, with awesome reverence. Though aware that the American tourist knows almost no Turkish, they will bombard him with hundreds of questions, all asked with the most beautiful smiling eyes. Content with no reply, they will continue, joking among themselves in the most affable and whimsical manner. For the Turks are the most friendly of people, eager to share their knowledge of their ignorance regarding the behavior of a jeep or a Cadillac. They will agree that though their roads are no good they are good mechanics; that Atatürk was a saint; that Turkish soldiers in Korea loved their American buddies; that Turks and Americans have a common goal regarding Russia and communism; and that the Hilton Hotel in Istanbul is something like a shrine.

When the foreigner takes off in his car, the crowd will disintegrate slowly; the scraggy, bewhiskered men reluctantly returning to work and the women to their washing in the public fountains. The children will go back to their cobblers benches or tinsmith employment, and the dozing policeman to his never-ending nap.

Not having seen a pig in all of Turkey, I nevertheless refrained from making inquiries about them, though I saw turkeys, with their fez-red throats, and heard their obviously "agglutinated" gobbling. The Turks call their turkeys "Hindies." But the Turkey-talk itself is purely Turkish, resembling the confusing word *yok*. The closer I came to the Black Sea, the greater it varied, going from *yuk* to a gobbling *yawk* and finally to *yoke* before I left Turkey near Mount Ararat.

But as the Turks went, so went I, when I discovered a hole in the seat of my old khaki pants. This, finally, was acceptance on the basic levels of life; and to celebrate myself as a completed Turk, I went to a tailor and had half a dozen colorful patches added to the seat of my pants. My ski jacket, which I wore to protect myself against the hot winds and the dry heat, looked even more interesting, Turkish-style. The pockets, long since worn and torn, were repaired with crazy-quilt patterns and colors. My shoes, somewhat battered, remained their American selves. But my scooter, always dusty and dirty, looked very Turkish, being streaked with brown, gray, red and white earth from the Turkish terra firma. I was a Turk, at least by my dress, but an American with a tight schedule, unable to pause too long; for my visa to enter Iran was soon to run out.

The heat became blistering near Havza. A few prairie dogs ran from their holes and dashed right back. A stork followed a peasant seeding the earth, its majestic stooping to pick up the seeds hardly noticed by the peasant. I became a stoolpigeon and told on the stork. The peasant hurled a stick, missing the stork but saving his future wheat. As a reward, he invited me to his house for a drink; an earthen dwelling set among low hills and dry plains, with plantings of thin Lombardy poplars embracing the village, but clustered with greater density about the village wells. Near the well a man was throwing water on two black bulls to cool them off. There were roses and larkspur in the fields. Behind the fields the snow-peaked Gümüsaciköy mountains loomed up. I smelled the Black Sea.

As I neared Samsun, a large Turkish town on the Black Sea, I crossed range after range of violently steep hills, with the Russian winds often hitting me with brutal, cold-snapping velocity. Now everything was contradictory; a few minutes of heat, at sea level, then the harsh hills and the winds rushing icily.

It was a strange encircling approach to the Black Sea, with no signs anywhere, just the looping road. Whenever I passed a peasant, I would call out for the name of the village. When he answered "yok," which did not help me, I would then ask, "Is this Terme, Ordu, Kavak or Merzifon?" The peasant, not wishing to be unobliging, would call back all the names I had just called out to him, so that I had an interesting game of place names but without a name to place myself accurately.

When some soldiers stopped me at Kavak to have a look at my *pasaport*, I learned I had been traveling through a restricted military zone since leaving Ankara. Was I to be arrested? Nobody knew, for nobody understood each other. One soldier lectured me in explosive Turkish and English, pointing to the nozzle of his rifle where the bullets came from. Next he pointed to my head and took out a dirty handkerchief. Apparently he was illustrating what would happen to a spy. I insisted that I was not a spy; that I was a free-wheeling American, and there was a difference. The soldier merely pronounced my Russian name, which almost made me a spy by the way he said it. He was still fingering my *pasaport* when I started up the motor, yanked it from his hand, yelling, "I will tell Atatürk on you, you pasaport!" and off I scootered, or so I thought. A few moments later a jeep filled with soldiers and officers was after me.

I was transported, or rather escorted, to see a major, who studied my passport as if it were a picture game. Had I trespassed willingly I was asked in English? I shook my head, for the question was a stickler and I preferred not to talk this one out. I managed, however, to come back to the question when the major ordered tea and cakes. Why should I willingly trespass over snow-covered mountains, dressed for the summer? Would not a spy go without a scooter, which attracted too much attention? Was not half of the village now outside looking at the scooter? Would it not have been better if I looked like a Turkoman on a mule, to be a successful spy? And what sort of a spy was I, not even able to talk Turkish? All spies went to language schools, I insisted; and all I could say in Turkish was *yok, pis mis yumurta,* and *ekmek;* or three words meaning "no," "hard-boiled eggs," and "bread." Were those good spy words?

Finally freed, with long and loud applause from the men and boys attached to my quickly organized claque outside the major's office, I made off, feeling like an ambassador of good cheer, with little boys chasing me down the road as I headed for the Zigana and Kop passes.

I swam in the Black Sea at Trabzon, once the ancient Greek capital of the empire of Trebizond. The Greeks had gone centuries ago, but the holy Russians, with ikons and czarist soldiers, came for a brief visit in 1916 to help take Turkey out of the Entente powers. Not far away, along the Black Sea coast, was Batum and Soviet oil, and Tiflis, where the student Stalin had turned from one weak church to the Bolshevik hierarchy and the Marxist catechism.

I looked from the Trabzon beaches toward Soviet Georgia, off in the sea mist and the haze, wondering how soon the Russians would start for Turkey; Turkey so weak, yet Turkey so strong with its generous people, but Turkey in between, unable to develop after the death of the Ottoman Empire; Turkey, for which I had developed a great feeling. And I stared at the mist, caught in a little game I was playing—could I get a visa to enter the Soviet Union?

It would take a month for a visa, providing they gave me one, which I doubted; for too much of my past marked me the permanent anticommunist, who had written for a variety of anticommunist publications. But I was playing a game: If I got in, would I get out? I would spy, of course; that was absolute. I would spy in the restaurants, as the workers ate *kasha* (cereal), *pirozhki* (meat balls), *chorni chleb* (black bread) and *schave* (spinach soup); and I would spy where the managerial comrades ate caviar, *boeuf Stroganov,* and *kapchunka*—the dried herring that was a delicacy for nondelicate bureaucrats. I would be the arch spy, tutored and unscootered, taking notes on the better vodka, Zubrovka, drunk by the elite . . . and it would all end in one enormous expose—in a proletarian cookbook for the American members of the Communist Party, still stewing in their rebellious juices about the Workers' Fatherland. I would be some spy! I ended this Russian revelry and my short course in spying and cooking by photographing the haze over Georgia. For a spy must use his camera. One never knew what a haze could turn into.

From Trabzon to the Zigana Pass I entered a world of colorful climaxes. I followed a deep river valley, with Swiss-like elevations, through seas of clouds, shivering my way through the Zigana Pass, wearing all my clothes, along the steep, curving mountain road. Bridges were out at many crossings, and the scooter was helped across by smiling, talkative road workers. Holes made roads and roads were deep ruts shooting wildly up and down, time effaced, shot away by mules and heavy trucks lying busted and broken on the sides. It was a world of unsameness, lost in itself in trapped mountains, with only a fearful, skyward pass hanging a rung below heaven. I looked back imaginatively to where the Black Sea quietly touched the beaches; where brightly striped herring boats made the caught herrings a still life of all the centuries along the Black Sea. Nets were drying there along the Black Sea beaches and old men in black hats, fingering turquoise beads, under trees, drinking tea, not waiting to die, urging their blood toward the new century the Turks were still making. And all the Turks were one; all the Turks were singing, sad people, content to be singing, and people with enormous affections.

I was in a Gothic-looking landscape, and all the names of towns and villages merged into one sound, with de-barked trees rolling in a rushing stream; the landscape of forest and foresters, thick and blooming trees; the axe and the two-man saw sounding in the wood. Dancers and zither players in quartets here bridged all static emotions in the grace and zest of the ecstatic *horon*. A little peasant girl came by in a marvelous headdress, like an elfin cloche, with green leaves and white flowers towering above her head; a hamadryad unconscious of civilization, coming from the enshrouding green woods and valleys. But at Vakfikebir, it had to be politics again; Cyprus again; and a strong laborer saying over strong Turkish tea, "We ought to take a lot of shovels to Cyprus and shovel it back to Turkey." There were little boys repairing fish nets, not knowing where Cyprus was—

whether it was earth, sea, man, or just nothing at all. Then a heavily veiled Moslem woman, riding in the sidecar of a motorcycle, completed a new cycle in progress, yet veiled and frightened of her body before men other than her husband.

A pass cuts through civilizations, and this was mountain, lumbering land, with men in overcoats and heavy turbans stacking pine logs under snow-covered sheds. Yellow azaleas banked the slopes, also alight with patches of orchid rhododendron. Waterfalls tumbled out of the mist from the adjacent hills, breaking through trees, appearing in the fog like tall feathers. At Urman Harvesi, halfway up the summit of Zigana Pass, the clouds darkened, ready for rain or snow, ready to sink the mud and earthen road and shut off transportation and communication to the civilization down at the next town.

It was getting dark and I was miles from a town, dragging slowly through the snowy, freezing pass, my weakened light hardly penetrating the darkness. When the battery finally gave out, I used my flashlight, holding it in my right hand along with the ever changing accelerator; for I was shifting from first into second every hundred yards through the up-and-down mountains. But it became much too dangerous, for my stiffened fingers ceased to respond. If I continued, I was inviting something; and I had gone almost five thousand miles without a serious accident. But I was cold and hungry; and the next town was miles below, in the darkness of the valley. I quit when I saw a grooved-out cave in the side of the mountain and pushed my scooter to the entrance to block off the cold wind slopping through.

My emergency rations had been used up frivolously, some given to children who had pushed me when I could not get started. I went hungry; then the damp cold got worse. There was no firewood and nothing to burn on the snowy mountain, for I was way above the timber line. I burned my special beat-off-the-dogs stick, then all of my matches, and I listened to the water dripping in the cave, imagining the walls would soon fall in and box me up for eternity.

Then I heard something, and I sweated. I pointed my flashlight at the scooter-blocked entrance. But I saw nothing—nothing that I could not hear better: the howling of wolves not too far off, the sickening wolf-blasting echoing through the cave.

Should I ride down to Koyans? Would the scooter really block out the pack? Then I remembered wolf tales my father used to tell me as a boy, and about the twelve years that he was in the czar's army in Siberia. I remembered, also, that a few days earlier a Turkish army officer had told me how one of his friends, stationed at a lonely outpost, had been chewed up alive the previous winter. Nor were wolves confined to the mountains, he assured me. They often roamed the streets of the most civilizied towns of eastern Turkey. I thought of Ilya, the Greek, bearing me a gift of a gun, which I had stupidly refused.

I started the motor of the scooter. It made noise, exploding a barrage of monoxide, filling up the cave. I stopped it and flashed my light; there on the rim of light the wolves sat and howled. I sweated into a mass of fears as I huddled into myself. Again I started up the motor, revving it up and exploding it like a machine-gun. I had found a weapon, one which could kill me with its fumes; a sound-maker, exploding its furies whenever I revved it up. But what if the spark plug got fouled or the fuel ran out? I had fuel enough for a hundred miles or four hours on a good road; but I was in a cave, not riding, just revving. I stood there, facing the scooter's exhaust outward toward the pack, thinking, praying, fumbling, freezing, soaked in sweat; and when the fumes became too strong, I cut the motor down for a few moments, looking toward the pack in the darkness, listening to their howling, then revving, revving revving, revving, all through the four-letter night.

By morning, or what passed for morning, I was maddened with talking to myself. I had made speeches to God, to nature, to the wolves. I talked of many intimate things, remembering my divorce, wondering about her . . . blaming her for this insane voyage to death; and I remembered my Paris good-by . . . and I knew all of my own sins, my transgressions and omissions . . . and why God had forsaken me. Was this the death I was to know and not to know?

But it was light; and the pack had gone, the snow tracks

pointing down the sides of the mountain. I looked at the tracks and watched where they headed. It was not toward Koyans, but in another direction.

I finished the last of my cognac and warmed up for a few minutes. Had I taken it during the night, I would have fallen asleep in the cold, and the motor would not have been revved. But most of my gas was gone. I coasted to Koyans, found a hotel, and slept for twelve, pack-haunted hours.

When I awoke, it was too late in the afternoon to think of going on and being caught out again at night; one pack of mountain wolves would make enough tales for my grandchildren. As for my hotel, it was really a sad café, without food, water or toilet. One fetched the water from the village well, and went elsewhere to be natural. But despite the complete lack of civilized plumbing, I decided to stay over another night. One could always use the windows, only there were no windows—just torn-out openings; probably the holes in the walls were what I had been seeking. But the room did have a bed. In any event, I had slept the sleep of the alive, feeling for my legs, my head, my arms, to make sure I was all of one piece and had not left a foot in a wolf's mouth.

Toward evening the "hotel" got busy. An Iranian doctor with seven children and his American wife from Staten Island descended on it; all of them piling out of a De Soto, much as if they were in the Catskills; all of them hungry and tired, with the children talking Farsi and English in one grand splurge of bilingualism. The wife, now more Persian than American, still managed to retain enough of her basic American democracy to chat about Staten Island, though all nine of the family wondered what I was doing so far away from New York. Before I could talk about the night of the wolves, the café owner began selling the Persian family bed space, though he had but one bed and one room left. It was a two-room, all-excluded hotel.

"And where will the children sleep?" asked the doctor.

"Two on the bed, two under the bed, and three on no bed." Then pointing to the doctor and his American wife, the proprietor added, "And you can both sleep in the car."

But there was still another passenger to be accounted for, a goateed German hitchhiker who had been stranded at the Iranian border for lack of a visa. The German was getting a lift to Ankara to get the visa. I quaked, for my own visa was now void, having expired between Ankara and Trabzon, and I had neglected to renew it at the Iranian embassy in Ankara.

But the German seemed indifferent, or perhaps he was a member of Garry Davis' passportless world. But I was not. The Persian doctor and his immense family bundled back into the car, laughing off my warning about wolves in the mountains. But he was a doctor, and doctors could prescribe for torn limbs, wolf gashings, and charge it. I went back to bed and heard the car speed off, heading for Zigana Pass—and the wolves.

The next day I was a wary traveler, my head filled with dreams of werewolves. I was Romulus and Remus in a Turkish setting, waiting for my real father to claim me—the wolf-child, Roskolenko, brought up on goat milk, which my wolf-father got regularly from a friendly goat, not wishing to eat the goat as long as it gave milk. Then one day papa wolf and I ate goat-on-the-spit, and the following day there was no milk. The next day there was no Roskolenko, for papa wolf got fed up with my sponging on him, and ate Roskolenko raw, with huge hunks of *chorni chleb;* for with black bread I was definitely more edible.

From a mountain tea-house I saw the steppes of Asia; short turf and plowed fields, and the high, snow-patched Soghanlis, ringing the tea-house on the plateau. It was bitter cold at the bare and desolate tea-house, which was too fantastic; much as if someone had put it up as a stage set to make the area more inviting. The few local customers ate soft-boiled eggs and ripe olives, digging them out of a collectivized dish, with tiny spoons.

Outside, a lone Mohammedan grave, with rough stones at the head and foot, increased the sense of desolation. But when a truck carrying high-piled lumber ran into the roof of the café and almost dismantled the top of the building, there was pandemonium, with all sorts of homemade legalisms suggested as a way out of the structural mess. One old gentlemen, whose face was more mahogany-burnished than human, found the inspired answer: that the driver repair the roof with his own lumber.

I didn't wait, for it would take more than the old man's legal genius to fix the roof. I went on to Bayburt, an extraordinarily picturesque town at the foot of the plateau, its houses set high around an ochre-colored cliff. A short distance away was the ragged wall of ancient Bayburt; then a green-appearing river in full tide, rushing alongside the brown, serrated town.

When I asked a man where I could get gas, I landed in the police station instead, where I got tea, then the *pasaport* business. Now I was no longer a spy; nor was anything said about my trespassing into forbidden areas. It was just an enforced social call, with tea, free gas, and large crowds examining me, the scooter, the free gas as it was being poured. They also helped me study my tattered map, which had the magical Mount Ararat lost between one of its many folds. When we finally found Mount Ararat, cheers went up and there was another tea for the road; but the road was no longer a road, just a mass of rocks, stones, sand and pebbles. As I was about to bump off, an imaginative old Turk rushed up to me, stuck out a torn czarist ruble note dated 1916, and asked if it was an American dollar. He had kept it for forty years, since the czarist invasion of Turkey—and finally a chance had come to dispose of it. I had to say that it was not even Russian; but if he got a small enough herring, he could wrap the herring in the ruble, or the ruble in the herring.

I had learned something in Turkey—the art of badinage, which I managed to incorporate into a system of communication despite my almost languageless American state. I joked at all times or mimed my way out of trying situations. I didn't need a gun, except with the wolves, for humor was a better weapon and a better friend. But then, this was Turkey, where even wolves could be talked out of their dining habits. And I talked to the police as if they were old friends, joking with them or giving commands that even surprised me. Whatever it was, I was an expert at it, and this got me over the human and natural bumps.

Scootering through the Kop mountain pass, before Maden, I was once again in the icy winds. My homemade, hot-weather havelock was now transformed into a hat with ear-laps; and though it was a day in June, I was soon wearing two shirts, two sweaters and two pairs of pants. My hands bulged with two pairs of heavy work-gloves by the time I reached the warmer lowlands around Askale. I was a Turkish study of odds and ends as I scootered past a Turkish army ordnance depot, where a young officer roared with laughter as he saw me, harum-scarum costumed, flying off in all directions. My shirts were out of my pants, for it was now very hot; and the second trousers were half off, unbuttoned where they should have been gentlemanly secured. Lieutenant Suat Soyusu, an English-speaking ordnance instructor, who wrote poetry between studying bombs and projectiles, stood out on the road convulsed in mad laughter and disbelief.

But I materialized, soon enough, emerging from my collection of mountain costumes as soon as he invited me to lunch at the officers' mess, where both of us drank too much vodka and ate too little food, which is always wrong. On the walls were hand-painted pin-ups and man-made, soldier-style verse, with a touch of indelicacy and ribald, Rabelaisian wit.

Lieutenant Soyusu's father was a general and a weight-lifter, which seemed consistent with bearing arms, as well as a poet and a champion fencer. When the Korean war broke out, every Turkish soldier and officer wanted to go to Korea, and the few thousands who got to Inchun made enviable records for bravery and bravado. But Lieutenant Soyusu, a home-kept instructor lost among the home troops, took up competitive foil fencing and won second honors in the Turkish army fencing competitions. Yet he felt a sense of shame for not getting to Korea, since the anticommunist implications of the Korean war had all of Turkey's moral support. If this was intelligent anticommunism or Western democracy in military action, the Turks wanted to be the soldiers in the field.

As I left the dining room, we passed through the recreation hall decorated with more garish paintings, some of them depicting episodes in the Crimean War, when Russia wanted to take over the holy places of Palestine and was seizing corners of Turkey in an imperialistic orgy. I remarked, grinning wickedly, that my grandfather, a Russian, had also been in the battle at Balaklava. "I'm not married but my wife is," Lieutenant Soyusu said quickly, ending my embarrassing, embattled association to my grandfather's little war.

Later, I followed the Firat River toward ancient, sand-blown Erzurum, which squats on a high plateau between the snow-capped Sakrak and Bingol mountains. Once it was called Theodosiopolis and was an Armenian religious center as well as a military and commercial city. Its streets are narrow, poky, and wind-blown, laden with the centuries, piled up with rickety shops of tinsmiths, silversmiths and shoesmiths, with six-year-old apprentices crowding every available inch, their little bodies fitting into every corner, much like the fixtures—a backdrop for the old ways of commerce and handicraft. The people sit and drink their little glasses of tea, spit out polly seeds, never looking at a watch or a clock, just at the mountains walling in Erzurum. The most solid section of the antique town are the hundred-odd jewelry shops that sell rings and earrings made with the obsidian-like black Erzurum stone.

But before I could enter the room, I was stopped by a jeep. Actually I had yelled out "Hi!" on seeing an American flag flying from the jeep. Out stepped Colonel Nelson, in charge of the American Military Field Team at Erzurum, and his translator, Mr. Sezai. They said, "For Christ's sake!" astonished at seeing an American. I would dine with them but would be housed at the Temelli Palace Hotel. And off went the jeep, but not the scooter, for right behind me was another jeep loaded down with Turkish military police. I was off to see the commissariat of police, to explain my presence in Erzurum.

This time it was very serious—I was in the heart of the fortified military zone, which was at least completely illegal. When I informed the commissariat that Colonel Nelson was my host, all I got was a Turkish *yok* and a *halvah* chuckle. "You will have only twelve hours here, escorted about all the time.

You'll be on your way out tomorrow morning at 5:00 A.M." I was on my way in a few minutes, heavily escorted, a scooting potentate, all the way to the hotel. My passport was taken—in case. When Colonel Nelson stopped by to invite me to a cocktail party, I told him that I was being expelled by morning; that I was utterly road-worn and needed a decent layover. Colonel Nelson phoned the military governor of Erzurum, and a forty-eight-hour respite was granted. I could relax, stand the scooter on its head, dance, cocktail, and prepare to meet Mount Ararat, less than a day away.

The cocktail party was return protocol—an adventure in American-Turkish social relations on the military level. I was the neat guest, in a pressed blue suit and white shirt, completely brassless among all the khaki on parade. I was introduced to Lieutenant General Necali Tacan, Commanding General of the Turkish Third Army; to Corps Commander Eendet Sunay, of the Ninth Corps, and to Division Commander Kemal Suer, or to more top brass than I had seen during the entire Pacific war. To get myself off pedestrian grounds, I talked about my spying operations; this became a severe test in both protocol and imagination. I said that I had been under partial military arrest all over Turkey. Did Turks hate off-beat travelers on scooters? I talked about donkeys again. Lieutenant General Tacan said that donkeys had made civilization possible, thus scoring heavily for my extra-legal brief on the cultural contribution of donkeys to both Eastern and Western civilization. The little, lonely yet lordly ass was ancient man's best cocktail friend and traveling companion, said Lieutenant General Tacan, grinning privately.

Three whiskies later, Lieutenant General Tacan, his colleagues and I were made honorary members of the Erzurum Rifles, Colonel Nelson's private hunting club, which used jeeps to hunt in the Bingöl mountains. While toasting to the new membership, a donkey was brought in by a Turkish soldier, having clattered up the one flight of stairs on pillow-padded feet. With much braying and donkey-like ceremonials, I was offered a gift of a donkey by Lieutenant General Tacan, and I became, for a few hilarious hours, the owner of my own ass.

The Erzurum Rifles was an all-inclusive, hard-working hunting organization, shooting at boredom and at animals; for the dozen American lieutenant colonels and majors were days away from socially active Ankara. They had two Turkish soldiers to cook and clean for them; a weekly courier who went to Ankara for mail, commissary and PX supplies; and their fishing tackle and hunting rifles. And since hunters hunt, as Lieutenant Colonel Busch said, "We shoot at everything that flies, swims, crawls, climbs or walks." It was open season in a closed-in, womanless world, and they shot beasts of the field and the air, including lynx, jackal, wild boar, leopard, hyena, tiger, crow, reptile, wildcat, marten, polecat, grey squirrel, otter, hare, fox, wild goat, partridge, quail, wild duck, yellow hammer, wild goose, woodcock, crane, great bustard, deer, wild sheep, kid, fawn, bat, hedgehog, francolin, turtle dove, hawk, wren, cuckoo, woodpecker, great grey shrike, female pheasant, wild hen, nightingale, swallow, stork, owl and starling.

Noting that the wolf was not on their mimeographed list, I asked Lieutenant Colonel Busch about this.

"And I thought this list was complete, damn it!" He corrected my copy to read, "and any and all the wolves intending to molest Harry, an American on a Vespa scooter, who must, by no means, be eaten up in Turkey."

On the forty-eighth hour, I was escorted out of Erzurum, along with Teheran-bound Point-Four executive Mrs. Lucy Adams, who had also trespassed. Three military jeeps took us to Pasinler, twenty-four miles away, where we were signed out of the military orbit and given over to Allah's world. I followed the wide Aras river over a brownstone bridge that had been built by Suleiman's architect, Sinan. It was striking, with its red-brown arches and turrets, a slender, stately bridge that man seldom used now, for the land was almost entirely desolate, dry, corroded and barren. Yet the bridge had use; it was the way to Mount Ararat, via Horason, which had a railroad, occasional trains, and even a semblance of modern, moth-eaten commerce.

But outside of Horasan, I was enveloped in a terrifying cloudburst; rain and hail poured from the heavens, the winds rising to a fury, making it impossible to drive or hold the scooter on the road. Twice great gusts hit my windscreen and threw me. I shot in the air and landed unhurt, with the wheels of the turned-over scooter cutting into the dirt. It became a problem of how to tackle the winds, of getting the winds to carom off my large windscreen and not turn over the scooter. When I started up again, I faced the screen into the wind and managed to keep upright, though I wandered into the desert, tacking as if I were on a sailboat. But by the time I reached the Sarikamis mountain pass, the storm ended and I could proceed safely up the twisting mountain and reach Agri, seventy miles from Noah's country.

I slept overnight at Agri, a town and name that stems from Biblical times; for much of this region retains variants of the names used during Noah's day. The Aras plain, with its red, rocky encrustations, is also called Archnoisda, or the "foot of Noah." The ancient Hebrews called the entire region Oardu; and Xenophon, when he marched back to Greece, called it Karduchi, or "Kurdistan."

In Agri, the next morning, I received an unusual invitation —to fish in the cold streams of Mount Ararat's foothills. An American-schooled Turkish engineer named Fethi Ulgen, who was working for an American firm building silos in the Agri region, was my host.

We left the hotel at 4:15 A.M., along with four other associate engineers and a Turkish army officer, in a jeep and station wagon bound for the high mountain village of Kara Seyit Ale. We went over open country, across donkey and goat trails, through mud villages and over mountain passes. The big rains of the day before had created a muddy morass of the hardened, sun-baked earth. As we forded one brown stream, the jeep got stuck fender-deep, and the station wagon, aided by six Kurdish peasants and a strong tow-line, pulled it free.

When we finally arrived at Kara Seyit Ale, the engineer, Fethi Ulgen, said solemnly, "This is Allah's Kurdish country and Noah's foothills."

By nine o'clock we were ready to fish, with Captain Mehmet Uludag from the Karaköse district the first man in the water. Actually, he was to be our teacher, for this was an Arab net-fishing experiment. We had neither hooks, poles, nor lines, just a circular net weighted on the outside rim with small lead balls. Captain Uludag gathered the net carefully in even folds; and when he cast it out, it opened in an oval shape and landed on the surface of the water in a complete circle, sinking to the bottom of the shallow stream almost immediately. A few seconds later he started hauling back on a black line attached to the under rim of the net, which closed the net and such objects as the net had scooped up.

The first throw brought back rocks. Three throws later Captain Uludag hauled in the first fish, which weighed a pound and a half and looked like a gaily speckled mountain trout. The Kurds called it *ali balik,* or red pointed fish, found only in shallow streams of high plateau country in intensely cold water. And this stream came from Mt. Ararat, less than thirty miles away.

We moved on, with handsome Kurdish boys following us, anxiously awaiting the promise of things to come, like chewing gum, which Guner Akidil, one of the engineers, had in his pockets. At the next stop another half a dozen fish were caught in a dozen throws. Two hours later and a mile farther upstream we netted thirty-nine beautiful *ali balik,* and called it a morning.

Back in the Kurdish village of Kara Seyit Ale we were invited to have lunch. A grizzled old Kurd took us to his stone house where we sat on native kilim rugs and at honey and yogurt, bread and goat cheese, as well as strong green peppers, small honey pears and apricots.

Outside, two green-eyed, red-haired Kurdish children were joyously sitting in the station wagon, fiddling with gears, lights and brakes. Inside, engineer Ulgen was saying that the local bears kidnapped little boy and girl shepherds, fed them, and brought them up like Romulus and Remus. I thought it was a fish story, if not a bear story; but he insisted it was the truth, which the Kurdish peasant attested to.

Having tasted fish that came from Mount Ararat's waters,

my impatience to go there finally gave way, and early in the afternoon I left for Dogubayazit, less than two hours away by scooter, though the spiritual distance encompasses thousands of years of Biblical history. Approaching Dogubayazit, one finally sees all of Mount Ararat, or Noah's Mountain, "the roof of the world," as the ancient Armenians once called the resting place of the Ark.

As I approached Mount Ararat, a double rainbow spread across the entire front, following a half-hour cloudburst, repeating yesterday's storm. The enormous arcs of Ararat's rainbows loomed against the black, overhanging sky, with tremendous patches of sun jutting over the two peaks. It was like a visitation, a spectacle as natural as it was Biblical; for the whole front was dropping with hail and rain, an abbreviated encore of Noah's forty days and forty nights of deluge.

Dogubayazit, known in Biblical times as Bayazit, is an army post, Turkish style, over which Mount Ararat hangs like a great white parachute, never descending. The smaller mountains nearby are like jagged battlements, bare and saber-toothed. Dogubayazit itself has a small lively main street and adobe huts, suggesting a transplanted Mexican village.

The army officers sit at tables in the shade of locust trees, drinking tea. The hot sun beats down upon Dogubayazit, twenty miles from the foothills of the icy peak of imperious Mount Ararat, which stares into the heavens as an eternal reminder of Noah's disembarkation.

I tried to get permission to climb it, but Colonel Resat Saran, my host at the Dogubayazit military garrison, said it was part of the forbidden military zone. Nevertheless, I was allowed to join some Kurdish peasants and their flocks a few thousand feet up the slopes of Big Mount Ararat.

While crossing the Aras plain by jeep, the Colonel's aide, Captain Ibrahim Eke, told me of his experiences when he attempted to climb Big Ararat. The summit, which he never reached, can, in good weather, be climbed from the eastern slope starting from a place called Bekare. But near the top, he encountered a sudden blizzard and sixty degrees below zero. When he tried the westerly approach he ran into a solid mass of impenetrable ice near the top. Unfortunately, he never got to see the alleged Garden of Eden, which supposedly hugs the foothills on the southerly side of Mount Ararat. "But anyone who manages to climb to the top," said Captain Eke, "goes by muleback to a town called Zeroba, twenty-four miles from Dogubayazit, to sign a book authenticating his successful attempt. But I have never been to Zeroba, unfortunately."

The peaks of Ararat are in Turkey, about forty miles from Soviet Armenia and Iran. The nearby Soviet town of Nakhichevan, on the Araxes river, translates from Armenian into "he descended here first." The entire geographical area is a vast, open repository of Biblical folklore, haunting the students of the Bible with awesome associations and implications.

Arghuri, the village that once nestled half way up Mount Ararat, means "the vine was planted here." In 1840, Arghuri, which stood near a great chasm, was devastated by an earthquake; but the vine, reputedly planted by Noah, is said still to be growing despite Arghuri's inundation with ashes and brimstone. The town of Marand, which I visited after Mount Ararat, and which means "the mother lies here," was the supposed burial site of Noah's wife.

Between the two Ararat peaks is a land mass twenty-eight miles in length, with many extinct craters evident. It has been suggested that Big Mount Ararat was once higher than Mount Everest, but that earthquakes shaved off thousands of feet from its height. Today it is but 16,946 feet, a huge, shouldering mass that has more "dome than cone" to it. Little Ararat is the handsomer of the two peaks, its elegant pyramidical shape 4,000 feet below Big Ararat, which received its name after the Armenian King, Arai, who ruled about 1750 B.C.

Mount Ararat was once part of an ancient volcanic region: Her first known eruption recorded occurred around 441 A.D., when darkness descended upon Mount Ararat for thirty days. Nicholas of Damacus, the historian, wrote, "In Armenia, above Minyas, there is a great mountain called Baris, upon which it is said that many who escaped at the time of the Flood were saved, and that one who was carried in an Ark came ashore on the top of it, and that the remains of the wood were preserved for a long time. This might be the man about whom Moses, the lawgiver of the Jews, wrote."

The Mother of the World has been, historically and theologically, the meeting place for brave, religious men, anxious to climb Mount Ararat and seek the Ark; and more than a dozen expeditions achieved success before 1900. Today, Kurdish peasants walk their flocks up the grassy slopes of the middle zone, going as high as 12,000 feet. The snow begins at 14,000 feet, which is about 5,000 feet higher than the usual snow line of the Alps, a phenomenon stemming from the intense aridity of most of the surrounding areas.

The first successful climber of Mount Ararat was a Russian-German professor named Frederick Parrot, who made it on his third attempt in 1829. The most recent successful climb was achieved by Turkish mountain troops during a week in August when the terrifying winds and snows do not lash the peak of Mount Ararat—the mountain "so lofty rising from the plains so low," as one poet called it.

Though Noah and his Biblical people once settled this land, there are no Jews today in Dogubayazit. Although it was once part of Armenia, few Armenians now live here; for this almost barren country is still in Biblical time agriculturally. The Kurdish peasant, patched and poverty-stricken, grows wheat and cereals and gazes on Mount Ararat, the conscious center of his life; for Noah is one of the Moslem saints.

"The little climb up the foothills calls for a celebration," said Colonel Saran when we returned. "If you can manage to come back in two months, I will get you permission to try for the top." With that, Colonel Saran made a toast: "To America and Turkey!" We downed an Ararat cocktail on that. Now Captain Eke toasted: "To retaining your male sex should you pass under a rainbow under Mount Ararat." I drank to that—then asked what Captain Eke meant.

"It means," said Captain Eke, as all the officers broke into laughter, "that according to folklore, should you pass under a rainbow under Mount Ararat, then your sex undergoes a complete transformation. We have a gentleman's agreement with our soldiers never to skirmish under Ararat when a rainbow comes up. It would cause a lot of confusion, you understand."

"Perhaps in Dogubayazit, but not in Copenhagen," I remarked. It was time for me to toast—and being Biblical-ridden, I said, "To Noah!" and a third mixture of strawberry liqueur, soda and vodka deluged us in the final toast, final until dinner, when the toasting began again, followed by *horon* dancing. Around midnight the party broke up.

A downpour started as they were taking me to the guest house, where Colonel Saran posted a scooter guard, informing the soldier that until I left I had the assumed rank of lieutenant colonel. "The only thing I can't offer you is good weather in Dogubayazit or on Mount Ararat," and he embraced me with a brotherly Turkish-style kiss.

"Of course, don't let this unseasonal weather fool you," said Captain Eke, continuing his Ararat tales. "Actually, it hasn't rained here forty days and forty nights for more than five thousand years."

I felt strangely purified the next day, watching the disappearing Mount Ararat; for it was magnetic, throwing me into a state of unusual but acceptable grace. I would have become a monk right then, or organized a private priesthood to celebrate Mount Ararat's relationship to birds, beasts and men. Here, on the frozen heights, one was at once so near to God yet so far away from man himself. I had reached the middle section of my journey; the man attuned, peculiarly; the man less lost, getting closer to himself, less the pagan and more the unconsciously religious man. But I would have no pictorial record of this place of grace. Most of my photos, unfortunately, had been overexposed.

I was once again escorted from a forbidden military zone—Mount Ararat. The year before, the Russians had complained that the mountain was being used as a spying window into Russia. I had spied out a poem, however, written months later, in Melbourne, when the remembered glow of Mount Ararat gave me my second experience in automatic writing. Once at Yaddo, I had given in to it—and I wrote a mystical sort of poem, dictated by the pressures of Hitler's murders. It streamed out, a hundred lines in five minutes, with not a word changed. When I showed it later to a beautiful artist named Marion, she remarked, "You look very pale, Harry. Have you seen a ghost?" I had seen myself—and I saw myself again when I wrote the poem to Mount Ararat. The poem is a failure, but it is interesting for its associated phenomena.

Captain Eke, my military escort, left me at the Iranian border, where, after an hour's wait, I was finally confronted by a fat, German-speaking frontier official who quickly noticed my expired visa. He shook his head and said, "Your visa is kaput, unfortunately. Three days ago I had to send back a West German to Ankara; but then, he didn't even have an expired visa. He didn't even have a passport."

"I met him and he looked very unhappy," I said.

"Ah, so. Are you unhappy?"

"I will be soon, I suspect. Those damned Turkish roads!" My protest looked genuine enough, but his reaction was still, "Ah, so. It is not so far to Ankara with that thing," and he looked at the scooter.

We talked for a long time. I was only an expired stamp, not a man without a passport. I had some smudged ink and a date—June 27, but today was June 28. I had been on the road for six weeks; I had talked my way out of Yugoslavia, where I had overstayed. But there was no use arguing any more. I would have to go back to Ankara—five grueling days on the scooter; with everything repeated, including the wolves. Seeing Captain Eke still outside the customs compound, I decided to ask him for permission to settle down at Mount Ararat—to become the old man of the mountain, whose universal visa had run out.

"Those damned Turkish roads!" I said aloud, starting back to my scooter. But my second bit of swearing performed a miracle. The Iranian official stuck out his hand for my passport and he soon stamped me in, and Ararat lost its future old man of the mountain, who would no longer have to sit out this peaceful century on Biblical ice.

"All mankind being brothers plighted," I whistled off, with Beethoven's "Ode to Joy" thrown to the hot desert waste that I immediately entered. This was Iran; and ahead was Maku, my first Persian town, straddling the sides of a red clay hill, its houses tucked under overhanging cliffs, suggesting a Persian variation of the New Mexico cliff dwellings. Long earthen walls closed in the houses, as well as lines of poplars; but the surrounding country was a veritable wasteland—plains of natural sculptured boulders, overpowering heat haze, and desert.

I was soon caught in the middle of a Middle Eastern debate when I searched for two empty gallon-cans which I wanted for reserve gasoline. A barefoot boy, who had more than cheek, offered to find them for me. Off he went, engaged in honorable findings; and off I went for tomatoes and gas. I found both on top of a hill street, the merchant selling me the gas at an unknown price, and only announcing it after he had filled my tank. It amounted to almost a dollar a gallon, though the legal price was sixty cents. I offered to siphon it back, getting out my little rubber hose during the debate; but I was already cooking with gas. He came down a dime but I still demurred. Just then the barefoot boy arrived and announced that he wanted the equivalent of a dollar-fifty per empty can. I told him, gently, that he would have to meet another American to get rich on. He refused seventy-five cents per can, saying that he would like to get rich on me, and he could find me a dozen cans.

A raggle-taggle policeman now joined the comedy, along with peddlers and their donkeys, their ears waving like Daumier judges as we haggled. I finally won out with the larger merchant; but the smaller one was adamant, the principled herioc type, though he offered to go down to a dollar a can if I bought a round dozen.

Three more policemen joined the Persian throng, each arguing for the kid's price. It looked like a riot now, not a carnival of buyers and sellers, with one donkey taking my side, nudging me and kicking at the small boy. His owner was forced to withdraw before things between donkeys, boys and men got worse. Having lost my donkey claque, I went back to telling off the cops. Was this the way to treat a stranger and a traveler? What did the Koran say—and I lashed out, calling them a mob of collecting thieves. I won.

Leaving Maku, I reached a plain of boulders and scootered through a black-walled canyon that ran along a narrow river. The country was not lush like Turkey, with little vegetation but many red clay hills. The Soviet border was a short distance off; and the town of Nagorni, between Soviet Armenia and Azerbaijan, not much farther. I remembered a Dr. Nagorni telling me about the town he was named after. He had treated me for measles, whooping cough, pimples, and assorted youthful complaints when I lived on Cherry Street, near the New York docks. He was a tender, heavily mustached man, who read many books. All the Russian classics were in his office. When I was but eight he gave me my first book, Dostoyevsky's *Crime and Punishment,* which I could not read for quite a few years. But whenever I came with my mother or my father, he would talk to them in Russian, which I then knew slightly, so that few medical secrets were kept from me. To get rid of my pimples I must get a woman, said the doctor jokingly. My mother blushed and my father thought that the good doctor had suddenly turned into an amoral scoundrel. What, a woman for a boy of eight? What would the little boy do with a grown woman? Apparently, I learned. . . .

Between Maku and Khoi, on the top of a mountain, was an Iranian police post and frontier patrol. The mountains billowed off in all directions, and the police galloped over the mountains as far as the Soviet border to observe their neighbors at their political beatitudes. Occasionally, a Soviet citizen would come across, usually walking with a pack of old clothes, having sneaked over from nearby Nakhichevan into Persian freedom.

The police fed me tea, though my unwashed glass had a collection of flies that somehow masqueraded as tea leaves. I was only nine miles from Khoi, the police assured me, cleaning out the flies for another glass of tea. Hours later, I discovered that it was seventy-five night scootering miles when I met a cyclist loaded down with camping equipment, a spare tire slung across his body, and a lean, hungry look in his eyes. Markus Dekker, a thirty-nine year old artist from Hanover, Germany, had been eighteen months coming and going between India and Germany, averaging fifty miles a day. He was skeleton thin, very esthetic in his looks, completely broke, hungry, fatigued, waterless, clothless—dressed in a pair of khaki shorts and a ragged open shirt to cover him in the desert sun. He made his way by making sketches, though I could not see enough strength in his body to lift a lead pencil. As for his customers, he had a few in India but none in Persia. Apparently the Persians did not love art as much as the Indians, according to Mr. Dekker. But I loved art and I gave him some of my emergency rations and my water.

Here was a tragic man—and I wondered what had started him on this reckless adventure. Did we have some similarity? Were both of us teched in the head? Was this the way to free the man inside by a new form of suicide? Whatever it was, here was artist Dekker and ex-poet or lost Roskolenko, finding himself; and we exchanged road data. As we shook a comradely good-by, Mr. Dekker said, with prophetic vision, "You have just entered a very difficult country, so you must be prepared for terrible times. Don't eat their food; don't drink their water . . ." and I expected him to add, "Don't breathe their air." I saw him pedal down the road toward Maku, wondering which of the many thieves in Maku would pose for him. Mr. Artist Markus Dekker, whoever he was, was no stranger to me. We were similar people in the Iranian haze; the outer symbols were the same as we bounced over the rugged tracks, heading eastward and westward.

I saw my first camel-train humped against the late sky; then young partridges in a grubby field. This was the Persian garden, in its deadly strange, silent world; for I had encountered but one bus on the road, one jeep, one artist on a bicycle, and now a few camels and drivers. But suddenly people began to appear, working with sickles in the fields, or piling sheaves into golden hives. My few Persian words, accumulated painfully, were waved away as I tried to find out where Khoi was. It was between this and that place, and hands pointed all over the horizon and into the ground. I felt lost and languageless for the first time, though I knew the significant words and phrases for toilet tissue, gasoline, meat, bread, a half-pint of oil, thank you, good-by, go to hell, you're-not-a-crook-but-you-will-be-one-soon-nor-are-you-a-saint-though-your-mother-was-obviously-a-good-woman. I had, until now, had little chance to use the nicest of these words, except in Maku, where I had used everything. There obviously was no traffic problem in Iran, though there was a linguistic one, nevertheless. I was just disappearing on an Iranian plateau, for Khoi seemed to have vanished.

Soon I was climbing again, averaging about nine miles an hour. The Iranian road now turned south and headed into the afterglow, rising higher and higher; though my informants, over flies, had said it was only a small mountain. When it got dark I used my headlights as well as my flashlight to make out the endless hairpin turns. In some places, the sides of the road had fallen away and I hippity-hopped by starlight, flashlight, and headlight for several hours. Finally the flashlight went dead; but there was Khoi.

A Persian village is a mixture of blues and greens, of brown walls, tall poplars and heat haze; and Khoi sprawled, wall on wall, in earthen architecture, high-walled houses and greenery. My hotel was more garden than rooms, with twenty beds ringing round the bubbling blue-tiled fountain, which was the dead-center for everything, from hurried ablutions to toilet solutions. my own room had only six beds, all of which I tried on the hour, yet ending up on the floor with all the bedding making an Arab bed to bed me down comfortably.

A Persian hotel is the second act of a thriller; and every guest is a character with an opium pipe and the Koran—in the thriller. In the morning two clerks arrived at my many bedded bedroom, carrying two ledgers and an abacus. They worked for a half-hour in silence, looking up occasionally between their bookkeeping to see if I was still present; and deciding that I was, they continued figuring out what I owed the hotel. I had eaten shish-kebab, spring onions and bread for dinner; my breakfast was tea, bread and honey; my six beds, which made difficult accounting, had twelve sheets and six blankets, all on the floor; but each sheet had to be paid for. With their hushed voices and the abacus summarizing my almost mad investment in Arabian comforts, I finally exited from the Persian comedy of hotel manners and scootered off toward more mysteries.

But the drama shifted to the scooter, now loosened from stern to handlebars. The rings and piston, after six thousand miles, were worn down. My spare parts included nuts and bolts, the odds and ends to fit odds and ends, but neither ring nor piston. And so I half-sped on, filled with artist Dekker's dire prophecies, on a busting-down scooter on the deserts and mountains of heat-trembling Iran.

By noon the heat became intense; my red havelock, once again revamped to fit the weather, now mostly mirages mirroring the bouncing heat of the adjacent desert. I forded many acequias and rushing *jubes,* where naked children were swimming; for the *jube* is the water system of Iran, flowing between the streets and the sidewalk in every village and town. In its contaminated waters is Iran's past and present. The peasants sit over it, pray over it, wash and urinate in it, and bathe their dishes and their livestock in it. Like the Koran, it is eternal and beyond the Westerner's comprehension; for since water flows, water is therefore continually purified, no matter if it is really a sewer, which most of the *jubes* soon become.

But Marand was a sewer of another color: a human sewer, and hardly the Biblical site of the resting place of Noah's wife. It looked brutalized in the high Persian sun; the slovenly dressed policemen viciously striking at pushing boys and men looking innocently at my scooter. When one of the policemen went after a boy who had, with wonder in his soft brown eyes, merely touched it, I raised two fists ready to beat the living Allah out of or into the cop. But somehow it all quieted down; the citizens muttering at their protectors; the scooter quite safe, if not sound; my anger justified, if sudden; my fears mostly mechanical, as the fluttering engine gasped slowly up the small hill outside of Marand.

An old man in an Astrakhan hat, his face weary with heat and fatigue, begged me for a lift as I began to climb the hill. But I had to turn him down; for the scooter was wheezing and rasping as it climbed. Halfway up, it stopped dead and started to roll down. I braked, got the engine going again, but the scooter could no longer climb. The motor raced, growling into the Persian silences; but it was an unmovable beast of machinery, minus two healthy rings and a pulsating piston, imprisoning me in the Persian limbo on the hill.

I rolled back and tinkered, changing the spark plug and refueling, looking at the oriental spaces; at the old man, still sitting under the chinar tree, wondering whether he was secretly happy. Out of the spaces a jeep dashed up the road. I called after it, seeing "United States Army" printed on its disappearing shanks. I called and called; for the jeep was almost to the top, ready to go over the rise. But it stopped, turned about and came down.

The driver, a tall, husky Brooklynite, Sergeant Anthony Bifora, carrying a carbine, was soon tinkering too. He was en route from lonely Rizaiyeh to bustling Tabriz for monthly supplies, a crap game, and other Army activities come pay-day at Tabriz.

An hour later, I was where I was, still in the mechanical doldrums. Sergeant Bifora, deciding to go on before some of his projected activities evaporated, finally gave up tinkering and said, "It's less than fifty miles to Tabriz; but if you can't make it by five o'clock, I'll send out a two-ton truck to pick you up. Anyway, Major Ritchie's directly behind me, so you won't be left to the wolves." And he patted his carbine, preparing to drive off. "I killed two of them last night."

I made several false starts, then stopped to wait for Major Ritchie, who jeeped up in an hour, his Texas twang singing across the desert, assuring me that he could haul me without a two-ton truck. But I still wanted to make Tabriz on my own scootering steam, even if it meant on a wheeze-by-wheeze basis. Besides, Major Ritchie already had two hitchhikers filling up the back of the jeep. One was a little Kurdish girl with long red-brown hair who had been abandoned in an alley at the age of twenty-two months, several weeks ago—a sad-eyed citizen of an orphanage, wistfully fingering a bag of Army rations with the cigarettes out and more chocolates added. Major Ritchie was also hauling an American lady-journalist, to whom he had given a lift to the Soviet frontier for a quick look inside. But it was not the sight of the American lady but the Kurdish child on the lady-journalist's lap, now suddenly involved in the American sense of emotional security, that made me think that I was still in good-hearted U.S.A. Major Ritchie beamed his new fatherhood, saying that he had lots of love and affection to give the pretty little orphan. And he had it, as large as Texas, all father and very happy about it.

"When you get to Tabriz, you'll be our permanent guest," said Major Ritchie, about to shove off. "You'll make it; but we'll go after you if don't get in before dark. And we have a motor pool in Tabriz, with some good Persian mechanics to strip the scooter down and give it a complete overhaul. If necessary, we'll make a new piston and the rings, to get you on the road. See you in Tabriz."

In Tabriz, hours later, I finally brought the ailing scooter to the motor pool, where Sergeant Bifora directed a mechanic to tear it down and check everything that moved. "It may have swallowed a small camel—anyway, breakdowns are not unusual here. Two months ago we kept a young German motorcyclist here for a month. His motor had burned out completely and a new one was being airmailed from Germany. We got up a little kitty to pay for it. The same goes for you, if you need the dough. Settle down and relax, fella."

I relaxed. I was thin, wiry, bronzed, still cosmically carefree, but very, very run-down. My speedometer read 6,518 miles, with Paris an illusion lost in the bubbling landscape of Iran. Ahead of me, however, was two thousand burning miles of Persian deserts before reaching the bandit wastes of Afghanistan, should I decide to go that way toward Australia; then the Northwest Frontier and the Khyber Pass into India. After that came Bombay, Bangalore, Madras and Ceylon, where I planned to catch a P&O ship for Perth, Australia; then another 2,200 miles across west and south Australia on the grim Nullarbor desert and plains before reaching Melbourne and the Olympic Games . . . and so I relaxed, drinking a long Tom Collins.

I wondered about many things, from poets on scooters running their own lyrical marathon, to things more athletic, like the Olympic Games and what it would be like coming back to Melbourne, a city that I loved. I had come this far without being killed or successfully robbed or eaten by wolves; but the warning of artist Dekker spoke of things to come. If I got through physical and mystical Iran, I would come through elsewhere; for Iran had more high peaks and deserts than any of the other countries I had come through thus far.

I had another Tom Collins, happy to luxuriate and be at ease while a mechanic repaired my scooter; a guest of the N.C.O. Club in another little America, but really hot, dusty, ancient Tabriz just below Soviet Armenia. Besides, a crap game was about to begin.

Chapter 5

HOURIS AND OPIUM

THE UNITED STATES Army Field Team No. 3, en masse, adopted me and my honorable troubles, though I was not exactly an orphan. My recent home had ranged over six thousand miles of man's great past, making me, by scooting osmosis, a natural enough contemporary for orphanages and army posts. But this was real mule and donkey country, complete with bubbling gardens, the Zoroastrian sun, the relics of Darius and Artaxerxes, ketchup and hamburgers, as well as the artifacts of American civilization superimposed, Army style, upon the Persian plains and mountains.

Twenty-eight sergeants acted as my over-all hosts, with some solid gentlemen in brass, like Major Ritchie, seeing to it that everything human was made available to me from Tabriz's bazaar world. When not at leisure, the sergeants instructed the motorized part of the Iranian army, teaching them how to move on wheels instead of on camels. They were the big brothers of the gears, helping to reorganize Omar Khayyam's blue gardens, currently stinking with decay and obsolescence. But all things, including the decay, made me feel at home, nevertheless. I joyously ate Spam instead of shish-kebab as I haltingly rediscovered the American language after six weeks of linguistic combats in five Romance and non-Romance languages and lands. I was at home, if somewhat awed by it all, yet conveniently relaxed when drinks were forced upon me, being totally American in this semi-colonial setting, but almost collapsing from fatigue.

That is, I would have collapsed; for my eyes were badly bloodshot; my hands were a mass of new and old blisters; my reflexes were way off, and what I needed only a doctor could prescribe for me. But Tabriz excited me with its fantastic past, and it acted like a tonic for a few days. It was older than Mohammed and older than Christ; for Zobeida, the wife of Harum ar-Rashid, had founded the city in 800 A.D., or so said an Army sergeant named Malin.

Sergeant Malin was a learned student of the jeep, being quite profound in things mechanical; but he also had a working knowledge of the linguistic methods and puns of James Joyce, as well as some Persian folklore. But after a few days of Tabriz's three hundred-odd mosques, including the Blue Mosque, I discovered that Sergeant Malin was not what the doctor had ordered for me. It was peace and not crap games, said Dr. Stewart, when I finally visited the American Presbyterian Hospital, or I would really collapse.

Some of the sergeants had driven in from their base in Kaligar, riding for two drowsing days over Persian deserts on their monthly trip to Tabriz. After a few days of fun and games, they would take the terrible road to Teheran, to load up at the large U. S. Army Commissary, ready for another month of dreary isolation at some Iranian army post, wondering whether the nearby Russians would come down like lions upon the Persian flocks they were shepherding.

They were a mixed group of career men. One sergeant had been through the Army Language School in Monterrey, learning Farsi or Persian. In his slow Southern accent, he said, "Why, I had to look up an atlas to find out where Persia was when they told me I was elected to study Farsi and go to Iran. But now, one year and a language later, I can tell the difference between Turkish and Farsi, and they speak Turkish in Tabriz . . . but I know enough Persian to shoot a good game of bull-crap and to get some lovin' in with them sandaled ladies down that brothel compound."

But Sergeant Malin, who had elected me for his buddy, because, "You're the only other man out here who ever heard of James Joyce," had a head full of steam and a penchant for multiple digressions, as well as enough comedy to keep Team No. 3 happy, at least for the duration of a crap game. Over a bottle of Scotch that he put before me, he lectured, zany-style, saying, "You're a rooter, rooter man, Mister scooter man. You know, when I was run down by a drunken Persian driver, I was worried how much I would have to fork over for being in an accident not of my own making. I kept thinking—what would Ernest Hemingway do if he fell out of plane and didn't have an old fish to fall back on? Man, that scooter of yours is a clod, never to be ridden again over the wastelands of the farcical Farsi. So you had better get yourself on the walking end of a camel and start buckin' for home. Take this bottle along, for a man never knows what he really knows. And the fact remains, for it defines a term as well—you had better stay here and eat buffalo meat instead of all them Persian kebabs. So let's face it, man, that scooter is a clod, and none of the effin' mechanics at the motor pool can put it together again. It might be better to make an electric shaver out of it and shave over-hairy camels than to scooter to India and Australia. And do you know that a water buffalo is water cooled and regular tit cows are air cooled? Few people are that scientific when it comes to titty culture, man. Anyway, I eat titty because it looks like meat when it's off a buffalo's bones and because it bypasses the Persian stringbeans. What, a war out here? You're a clod, man, a clod in the sodomy of Iran. Get yourself a jeep with five wheels, using one wheel to steer with; or better, join the U. S. Army. In twenty years, if you're still alive, you can save enough to buy yourself a Cadillac. Anyway, in event of an emergency, and war is an emergency, I'm going to get me a herd of goats and become a Kurd in the hills. No, not a turd—but a goddamned Allah Kurd, so help me God!"

But it was pay-day at Tabriz, and Sergeant Malin had also driven for two days from Kalipar to collect his team's pay at bustling and exotic Tabriz. Tabriz had everything, including a decent cook who managed to cook for twenty-eight sergeants and make them love his Armenian cookery; but then, Tabriz also had a bazaar as well as dark, strangely exciting women; Tabriz had sunny streets and romping, unveiled Armenian and Persian girls, who had read enough about feminine freedom in the democratic West to desire it, and without veils; and, above all, Tabriz had an excellent sergeants' mess, with a bar to match. The mess had a fine green table that looked even better when dice rolled across its green fields and into the gambler's imagination. And Sergeant Malin was a fanciful soldier with the dice, as I soon discovered.

"It just takes one shake to make a point," said Sergeant Crown, the tall, thin soldier holding a double Scotch.

"It takes luck, science, art and prayer," said mustached Sergeant Banning, who wore Bermuda shorts after hours. "And you happen to miss out on all four, Malin."

"After you, there's only heaven, Banning," sneered Malin. "I'm shooting ten bucks, gentlemen."

"The philosopher has spoken, and I'm fading," said Banning, dropping ten dollars on the table. "Let's see how cold your war is, Malin."

Sergeant Malin rolled indifferently, making a pair of fours. Looking somewhat askance, he said dramatically, "Now take Major Ritchie. He doesn't smoke, he doesn't drink, but he adopts little Kurdish babies. I reckon he must have four at home now; and without even a wife to look after them, too."

"But he's got two hired maids, Malin," said Banning, calling for a double Scotch. "Keep 'em rolling, Malin."

Malin kept them rolling, but he made nothing to excite anybody. They were just dice splashing across the green table.

"And take that colonel-man over all of us," said Sergeant Crown, who looked even taller when he kneeled. "He's a real southern type gentleman, suh."

"You and he both," said Malin. "Any day now and he's going to become a Persian gentleman, after a farcical fashion. Bali! Come on, you dice, say a nice Persian *yes* to me. Bali!"

The dice said *no,* making a six and a one, and Malin started to swear. "That colonel, gentlemen, he don't like what I said, I suppose. From now on I'm talking like a philosopher. Shall I hold the dice or does someone else want to shoot? Okay, I'll keep 'em rolling. Shoot twenty, you fantastic soldiers of misfortune. Who's fading?"

Banning took it all, and looked at his drink. "I need another to go with the other. Hey, Joe, fetch 'em quick."

Joe, the Armenian cook who doubled for everything after cooking, fetched it for Banning.

"Bali, sir, bali!" said Banning. "Shoot, Master Sergeant Malin. The whites of your eyes are showing."

"The trouble is, the cold war's getting chilly," answered Malin. "Gentlemen, as of now, you're all about to become financially extinct. Come on, bali!"

Malin rolled again and made a seven.

"It looks like a clear victory for me, gentlemen," he said, refusing to pick up his winnings. "I'm shooting the forty, gentlemen. After this, Nirvana."

Banning faded half and Crown took the other half. The Texan, Bizer, was waiting to break in on his own terms, which meant taking five-dollar side-bets when they were offered.

"Major Ritchie has a great future in back of him," said Malin, rolling the dice again. "He can have a whole family and not even bother to have a wife. Now, that's perfect human economy. Come on, bali." Malin rolled a nine.

"That's dice economy," snapped Banning, not liking the point. "You can do much worse, I'm sure."

Malin, who seemed to have things on his mind, interrupted his shooting to take a paper out of his pocket. He spread it out for all to see, then began to declaim, "Comrades, in America the soldiers are badly paid, for they don't believe in dialectical materialism; and without that, how can you make money? How can you make anything at all, comrades, especially babies?

Think of poor Major Ritchie, who's opened a private orphanage to accommodate his fathering instincts—think!" and Malin rolled his nine.

He bowed like a footman, looked disdainfully at the eighty dollars on the table, then said, "Gentlemen warriors of the leading capitalist nation, may I shoot the works?"

"You may," answered Banning. "I'll take twenty of it. Are you coming in, Bizer?"

Bizer came in reluctantly, along with Joe the Armenian cook, and Crown.

"Put that damn paper away, Malin," said Crown, looking annoyed. "This is a decent crap game, not a bull session."

"I've never noticed the acute distinction, gentlemen," replied Malin.

"Then start noticing it now," said Banning. "Bull is just plain fodder, and crap is money in the wallet."

"If you make it, you'll spend it in the compound; where the girls with tight Persian trousers soon have loose Persian panties," answered Malin. He rolled, looked bored, hardly bothering to look at the dice after they hit the other end of the green table.

"Seven again, it seems," said Bizer. "I should have stayed out of this roll and I'd still have the twenty."

"Another materialist!" shouted Malin. "Think of your soul, your spirit, your sacred, untouchable pay, Bizer! Have you no financial beauty in your Texan soul?"

"You going to shoot the lot, the whole hundred and sixty?" asked Banning, hitching up his Bermuda shorts, now wobbling about his bony knees.

"I have faith, gentlemen. I have spiritual guidance and other nice things," said Malin, waiting for the fade. "Gentlemen, the cold war is below zero. Fade me and warm my soul, gentlemen. How about you, Mister scooter, rooter man?"

"I've just got enough to take me around the world," I answered, lost in the banter.

The bettors came in from all sides, fluttering to the table like pigeons mounting a branch.

"You take the temporary wife, the *sigheh*," said Malin, preparing to roll again. "She has absolutely no future in the Iranian Weltanschauung. Her cosmos is circumspect if not circumscribed. She is the low daughter for high dilly-dalliance, subject to the mullah, to her father, to the Koran, and to the top banana man; whereas, in New York, where career girls doth grow, there we have a unique social condition. The men wear bras and the women wear jock-straps; and this is the very latest thesis, written by an undergraduate from Lockjaw University. Bali, my cubes; bali me good!" and Malin rolled again. It was a five.

"Things are getting difficult in the dialectical domain," sneered Banning, trying to exercise his will over things animate and inanimate. The dice, to Banning, were animate, but only in Malin's hands. "Things are getting veddy difficult, Malin."

Five hours later, things were still getting more difficult for all but Malin, who had won a thousand dollars of other people's money. He was now drinking straight from the bottle, but not enough to unbalance his gambling values. He was saying, "Why, I'm so rich my head's bustin' out with gold bricks. I'm in full tilt, heading for a Yankee bank. But I lack a sense of geography, gentlemen; and that's what comes from going Farsi, rolling those plush-lined ivory cubes down the running *jubes*."

"Shut up and start rolling in the *jubes!*" yelled Banning, feeling the pinch of poverty. Half of his pay was gone, with Malin possessed and possessing half of what had been the private property of ten sergeants, their monthly pay, in easy installment payments.

To me it was a drama, but a mad game filled with all sorts of private meanings. I never gambled, but I was enjoying this spectacle fully, wondering what Malin would do when he won. Other sergeants had joined the game and it was now a free-for-all, with a few intoxicated soldiers standing around to kid the game.

Malin started rolling again; and it was another seven, looking straight up with sarcastic effects. Under the gleam of the blue daylight bulb the dice really looked possessed. Outside, the lamenting radio music had begun; a Persian woman was singing the usual dissonance of sadness; but the whole of Tabriz seemed to come to a stop with every roll of Malin's magical dice. And though the dice had changed hands dozens of times during the eight hours of play, at 9:00 P.M. they were again in Malin's hands, rolling toward his second thousand.

By 10:00 P.M. Malin's fellow-sergeants were broke and the game was over. With typical generosity, Malin asked, "Anybody like a loan to tide him over for about a month in Iran? Happy to oblige, gentlemen." Malin finished his bottle and ordered Turkish coffee, "At least a pint of it, 'cause I'm going to the whoring compound. Anybody care to join me or make a little loan, gentlemen of misfortune? And Mister scooter man, you're elected to come to the compound with me and I'll get you a real Persian lady to sleep with." Malin went off to the toilet, leaving all his money on the green table, neatly stacked. When he returned the stack looked even neater and larger.

Banning had borrowed a hundred, and Crown took another. The rest sulked, soldier-style, thinking how prudent they would be with next month's pay.

Finishing his Turkish coffee, Malin left his colleagues and wobbled out to his jeep parked outside the mess. I followed reluctantly. Malin was now very, very drunk.

Two Persian boys were hanging around the jeep, and they begged Malin for baksheesh. He tossed some rials at them said, "Now, you Persians without a garden to piss in, get thee behind me," and they dashed off, hugging their dime's worth of rials like sudden riches.

"Get in, Mister quiescent Harry." I got in. I wanted to see the compound; to see what would happen to Malin and his money.

Malin drove soberly until he reached Tabriz's main street, which was still crowded with peasants, bundle carriers, droshkies and cyclists. But Malin's driving was off tonight, for he almost cut off the head of a man coming up with an earthen jar from a manhole in the center of the street. The man ducked in time, dropping his earthen jar and breaking it.

"Maybe I can still get him, Harry," muttered Malin. He stopped, reversed, returning to the peasant once again emerging from the manhole. When the peasant saw Malin, he let up a wail, and ducked again, thinking that Malin had come back to finish him off. Instead, with a gracious bow, Malin said, "You hauler of water, you poverty-stricken clod of Iran, I give thee a buck in the form of a Yankee dollar; so next time keep your bloody bali head deep down that bloody *jube*," and Malin dropped a dollar into the peasant's wavering hands. With that, Malin was off again, heading for the compound.

It was difficult to find; for we got lost near the bazaar, now dark with night. Inside the bazaar whole families lived out their days, nights and years, peddling to customers or waiting for customers; and when they emerged, they blinked at the sunny outside and went right back into the bazaar. The bazaar had everything that an Iranian needed, including women, and other sinful pleasures. It had restaurants, cafés, hot baths, opium. "But it always makes me sad," said Malin, "It reminds me of my boyhood in a dark, black, cold Pennsylvania mine."

"That bali blackness," he mumbled a minute later, finally finding the street of the whores. "We'll have a double-barreled orgy tonight, Harry, with water pipes, opium pipes, ladies with and without pipes, then Kalipar and purity for another month. Maybe I'd better get myself a *sigheh,* a temporary wife, and be good to her for the rest of my enlistment. But tonight I'll get me a dozen women and the same amount for you, Mister scooter man."

"One good one will be good enough," I said, wondering what I had let myself in for. I was the silent man now, and not very strong at the moment. I had drunk too much, and what I needed was sleep, and not with a whore.

Malin had talked about a girl back home, "Someone I wanted to marry, but it didn't work out. Besides, I'm much too sensitive for marriage with just one girl. I like them in bunches now . . . and a married man is not always a happy one," he sang. "And so I do my temporary romancing at the compound, though it's beginning to have aspects of permanence after a few years here. Maybe I'll make a break from it . . . if I do something strange, something I've always wanted to do."

I wondered what he meant, but I did not ask; besides, anything that Malin said was strange enough, normally.

We parked the jeep and started walking down the cobblestoned alley, with Malin bouncing as if he was in a trance. The place looked familiar; but all of them did, in so many Persian and Turkish cities, selling but one thing—women. But in Tabriz the girls gave out free tea and vodka as an inducement . . . and I laughed, for I saw a girl come out of a walled-in house carrying two glasses of vodka. Soon we were following her inside.

We sat on a patio and listened to four musicians playing Persian jazz, with the women, for there must have been over twenty rooms, coming and going between them. Love and sex came at you from every direction, without a stop for your yes or your no. It came out of the walls, from the music, from the drinks being served; and when one of the musicians wobbled over with a opium pipe, Malin grabbed it and smoked it quickly, with the awkwardness of an amateur overanxious to see blue dreams. "I'm going to the bali dogs, you musical infants! At thirty-two, my private war is either getting hotter or colder," said Malin, handing back the smoked pipe.

A girl in red pantaloons, wearing green beads and copper armlets, got us the usual shish-kebab wrapped in *nuun,* a dark, thin bread. Soon she came back with glasses and vodka; then the musician came back with his opium pipe; then two more girls, more nude than dressed, joined us, ready to take us to their busy mattresses. But Malin started to dance, doing a Polish polka by himself, which amused the musicians, if not the three girls concerned with other forms of motion.

"You like to dance with me?" asked the girl in the red pantaloons. "I have very gentle hands. I am like a professional dancer, sir, because I was taught by a man," and she moved sinuously, like an imagined houri.

"You are very professional," and Malin bumped back. "Everybody in the bali world is a professional about something."

One of the girls came up to me. I took a glass of vodka and said I'd dance later or look in at her mattress later. She was fat, round, big and overwhelming.

"I have a very sick mother; also a very sick brother," said Malin's girl. "My whole family, they all need doctors very much."

"Do you need a doctor, too? What kind of a sickness do you have? Do you have a sickness that money can make disappear?" asked Malin.

"Money make everything disappear."

"Especially mad money," laughed Malin. "I want another pipe. Get me more to drink, and get me more girls. I want all the girls in this joint, but not you. Tonight the whole world is going to disappear into girls. Hey, Harry, take your whore to the mattress."

Malin's girl hurried after more opium and more girls, wondering if the drunken sergeant would not disappear before she returned. She sighed, knocked on doors, then listened for the sounds of love-making. Hearing nothing, she called out, "Hurry, hurry! Come out in a little minute!" and she went on repeating her call to an orgy. When she finished with her selection of girls, she hurried after the opium, to the other end of the compound. In a windowless room was the preparer; the thin, aged man who assembled the pipe and the opium. And as if reading her mind, he had one ready, which he stuck into her hands, saying, "Take it quickly." She was off again, racing back to Malin, who had taken off his shoes and was lying back on a red pillow.

Malin sucked at the pipe a few times, then said, "Everybody is so sad. Bali. Are you sad, Harry? Fornicate, man—take that fat one to Nirvana."

It was going to be that kind of a night, pipes, women, sadness, only I wasn't sad. I was a tired man from a scooter, not from Malin's pressures of the moment.

"I'm relaxing, like the doctor said," I said, pulling the

fat girl down to my pillow. "Nature gave this girl everything in abundance."

"Everybody is so sad," said the girl in the red pantaloons. "There is nothing for me any more—just vodka and opium. But once I was a *sigheh,* living honorably as a temporary wife with a rich man. Then the man die, and I have his child but not his money, unfortunately. After that, no other man want me for a temporary wife. Mister soldier, do you want me for a wife? I will not cost you very much."

"What is a *sigheh?*" I asked. "Is she a concubine?"

"Everybody is so sad," lamented Malin, ignoring my question. He pulled the pantalooned girl to him and kissed her lewdly. Her response was equally lewd, and Malin laughed and laughed, his hands going after the buttons on her red pantaloons. But the girl jumped away, shouting, "There are too many, many people here. Better that I dance," and she whirled and wailed, her voice echoing with lamentations. But the sadder her voice grew the more riotous became her hip movements, with the musicians following her antics in loud musical accents.

Now Malin danced; then the girls from the busy mattresses, who had finally emerged. It was a whirling, profane spectacle, the patio looking as if every whore in Tabriz had suddenly become a dancer for Allah's sad citizens of sin. Malin was now terribly drunk, unable to stand up straight, and he was shouting, stomping, falling then standing up, his voice mounting, as if this was his last call before he lost himself in his purge. "I am having my private Point Four, ladies . . . line up . . . get your wealth now . . . get it before I change my mind . . . or before my mind leaves me for good. Tomorrow the world is going to disappear; the bomb in myself is going off now; and we who are about to commit fornication, let us salute each other . . . for God is love . . . and I love all of you . . . I love everybody . . ." He fell, finally. I ran to pick him up, but he was surrounded by hordes of women, billowing and pillowing him with flesh; all but the girl in the red pantaloons, who said, quietly, that all Malin needed was one woman, a temporary wife, and not all those disgusting whores.

"Point Four, get thee behind me!" sang Malin, finally doing something strange. He fished in his pockets, as if completely in control of his senses and his movements. His own pay he put into his back pockets; but his winnings were soon flying about the patio, tossed in all directions, with the girls and the musicians and the opium preparer leaping about insanely, unable to understand the denominations they were picking up. They were old, green, American bills, so familiar yet strange; and they rushed like whirling dervishes toward the fluttering bills.

I watched, drunk and stupefied, unable to understand it for a moment. So this was Malin's purge—four thousand dollars from a crap-game—thrown to the winds. I wanted to join, to make for the bigger denominations; but I would have lost face or something. Besides, it was Malin's strange party, and I was a guest with a very fat whore at my side. She had tried to jump in and grab for the bills, but her lack of agility and her weight kept her in poverty.

Nor had the girl in the red pantaloons joined in. She stood near the room where the opium came from; and Malin was stumbling toward her, meeting her at the door. He had executed some sort of a strange wish, finishing something that enriched him for a minute. The women along with the four musicians had his cash, from some orgy for charity that he had indulged, Oriental style, to satisfy a whim fourteen years old. But now it was all over; the act was completed, and the man was facing himself in his moment of imagined greatness.

The girl kept Malin from falling again. A moment later she was leading him into the opium room, where her hands were probably gentler than they had been since her days as a *sigheh*.

When Malin arrived back late the next day, he had all of his own pay still intact. He looked strangely moral, too, as he told me what had happened. "She refused every effin' cent; would take nothing, man. And what happened to you, Harry? Did you finally have yourself a time, man?"

My head ached, but I'd had myself a time. I'd had a purge,

too, but it didn't cost four thousand dollars—perhaps only eight tumans or about a dollar.

"Was that all the fat girl wanted, Harry? And how long is the bloody Liffey river? Anna Liffey Whoreabelle—she flows fifty elfin miles; from the mountains of Wicklow to the bali Dublin Bay. Did you have yourself a time, Harry?"

Sergeant Banning, looking very mysterious, came over with a double Scotch. His shorts bulged peculiarly. In back was Sergeant Bizer followed by Sergeant Crown, unable to contain themselves, laughing with mad, uproarious gusts. Behind them followed all the other sergeants who had lost to Sergeant Malin, and they just looked, bubbling with mirth and mystery.

"Gentlemen, a man has a right to dance at his own funeral, if he can manage to arrange that phenomena—and I did that last night," said Sergeant Malin. "If any of you think that I ought to see a head-shrinker, let him start thinking out loud."

Sergeant Banning put his drink down and dug into his bulging pockets. He took out two wads of bills and laid them on the green table.

"I'll make it very simple, Mr. Liffey James Joyce," said Sergeant Banning. "In one pile is a lot of stage money, sent here in answer to an ad in some screwy stateside men's magazine; in the other pile, Mr. Liffey River Malin, is real money—four thousand bucks in real green silk paper—and them bucks belong to you. We pulled a fast one on you when you went to the head and left your winnings on the table, too drunk to see the difference. And now, how much do we owe you, Sergeant Malin?"

"Nothing, gentlemen, nothing," said much sobered Sergeant Malin, refusing to look at the two piles. "Nothing at all, gentlemen. Better still, everybody take half of what he lost out of the right pile—and if he don't, I'm never going to talk to that Liffy bastard again or win money from the clod."

"Well, since tomorrow's July fourth, we'll consider it part of the national gain," replied Sergeant Banning. "Gentlemen, line up and take an honest count; only half of what you lost yesterday, and not all year. Thank you, Sergeant Malin."

It was just like that. But it was Iran, where strangeness was normal and soldiers became strange—adopting a protective coloration for Persian hallucinations.

It was a real Fourth of July celebration, and American Consul Dresser invited all the local Americans along with the local French, as well as four silent Russians attached to the Soviet Economic Mission. Invitations were also extended to the head of the Tabriz Secret Police, assorted businessmen, Persian politicians and unsocial socialites. The Russians, dressed like successful proletarians gone bourgeois, with the proceeds of economic determinism showing in their left eyes, were completely one in their self-sought exile in the gardens of the consulate. They were impossible to talk to or be with; improbable and peasant-like in their gaping stolidity, and quite mystifying in their drinking habits; for they drank beer, which they had been told was Persian and very poor, and not the vodka which some wit, probably Sergeant Malin, had decorated with small American flags.

After due Fourth of July pomp, but without ceremonies, the food came, including popcorn and salmon hors d'ouvres, everything colliding in a quaint army-style melange. Colonel Stapleton, head of the American team, and Consul Dresser beamed pleasant protocol over all as they engaged their foreign guests in hesitant bits of allowed tête-a-tête consistant with the rigid forms of official repartee. The army sergeants, however, lapped up the liquid entertainment as they staged their horseplay in a not-too-discreet corner of the overflowing garden, with Sergeant Malin, who had elected himself as lead-man, ably seconded by Sergeant Bifora, prodding on Sergeant Malin on to greater flights of Malinesque fancies.

"Why, I'm the only proletarian in this bloody garden," said Sergeant Malin, pointing to the fattest of the four silent Russians. "I went to work in a Pennsylvania coal mine when I was thirteen, replacing a mule that had just died from human malnutrition. See my revolutionary palms, men of arms! Ask one of them Russkies to show their collective callouses, and you'd find them on their bourgeois bottoms. Economic Mission my arse! I beg your pardon, Sergeant Bifora, but I really mean my bottom, as the boys around Piccadilly Circus used to say to the girls around Piccadilly. Ah, my dear Tovarischi; you dolorous proletarians in capitalistic uniforms—" at which point Major Ritchie soon put an end to the comedy and sent Sergeant Malin off to his jeep, with orders to drive back to the N.C.O. Club and sleep off his July fourth and his July third digressions and transgressions. But while driving back, Sergeant Malin once again went over the same manhole, almost taking off part of an ear from the head of a man, who ducked before he was completely guillotined.

These holes were the only source of fresh water for the poor, and everyone made at least one daily pilgrimage with his copper or earthen pots. If cautious, the adventurer in search of water ducked the diverse droshkies and toxic taxies rushing madly through Tabriz. If he was a porter, or a *kamal,* he gave up his heavy burdens for a few minutes of respite around the yellow waters in the *jube,* saying, *Insallah,* "if God wills," as he offered up his Moslem thanks for his pock-marked, tragic life. And the poor woman washing vegetables alongside of the porter intoned sadly, offering phrases to good health and good fortune. She herself would receive little of both, but she happily offered it to a woman busily engaged in bathing a filthy, half-blind child in the *jube.* The workmen coming in from the nearby brick kilns or from the fields would say that things were all right, as they stripped to their waists to bathe in this over-all communal well-spring of Iran's sanitation system, which really served up disease, not life, from its fountainhead. It was a flowing sewer, mostly, a trench for anything and everything, a ditch between the sidewalk and the gutter, with the supposedly hallowed waters making the people of Tabriz permanently sick.

Tabriz is a huge, colorful bazaar, with miles of glass-domed alleys emblazoning the variety of its handicrafts. It is an open city during the day and a shuttered mystical entity at night.

The shops, which are tiny affairs lost in the labyrinths and endless dead ends of the bazaar, are the social-cultural repositories of this ancient city. Whole families work and live there, sleeping on rugs or benches, using old coats and shoes for pillows and blankets. Here thousands of children learn their trades and their futures, seldom leaving the dark catacombs of the dusty, wealth-laden bazaars that sell everything needed to sustain, if not to beautify, the contemporary Persians of Tabriz.

It is all bustle and noise. The tiny tea-boys dash about, and the bread sellers take their time dispensing the thin, tasty *nuun*. The rope seller is more patient than the diamond dealer, since everybody needs a rope in Tabriz. There are camels and mules and baggage, with ropes attending everything, for the rope hauls and holds Iran. But the adjacent man blowing his bellows is a very nervous type, the very blowing exciting him to blustering comment regarding his immediate neighbor, who is patiently uncoiling some hemp for inspection.

Overhead, the large encircling glass domes, which hardly let in light, have pigeons swirling below them, adding to the leafy curtain that appears to enshroud the interior of the bazaar. It is a man-made heaven, decorated with the blue tiling that Persians love. The blue is as radiant as the deep, hazy, sun-baked brown is all-consuming in its drabness.

At the wool market three men are smoking one hubbly-bubbly pipe while waiting for customers. Each man takes his turn, then he stares ahead into Persian space. At the nearby carpet halls, which are huge caverns piled high with rugs, there are customers arguing over Kerman rugs. To soften the process, the owner serves tea, placating the buyers, two Armenian men dressed in European-style business suits.

Twenty feet from the rug market is a Turkish bath bubbling over in a palatial cellar. The waiting room is a blue-tiled garden, with the pretty fountain in the center used for last-minute feet washing. The bath is the heart of the bazaar, the cultural provider dispensing rest, soap, hot water, gossip and business deals.

A boy with a pet goat accidentally stumbles down the stairs and falls with the goat into the fountain. Embarrassed for himself and his pet, he is laughed out of the bath and kicked up the stairs, where he runs into a Kurd leading a rug-laden mule to the rug market. Now the goat finds itself entangled in the mule's harness and rugs, and once again the boy is beset for his pains. Next he stumbles into a blind man yelling "Baksheesh" and he gets a wallop across his back with a cane. Scurrying off, he loses one of his shoes, which are bent back at the heels. He retrieves his shoe, then dashes with his goat to the nearest exit, having had more than enough of the bazaar's bizarre ways.

It is also a place for spices; and sacks of saffron, root, dried lavender and rose flowers melt into the gloomy areas, along with cinnamon sticks, dried orange peels, cloves and curries. Near the spices are trays of pigs' feet as well as sheep hides and entrails; for the bargainers and buyers are many, each seeking the most for a few rials.

In this Gothic-arched darkness the relationships are tremendously striking, balancing off time present with time past; the garish and the conservative shouting away the commercial or synthetic values of Tabriz's blue and brown droshky world. It has taken the twenty-eight American sergeants in its stride, having known a variety of conquerors who came with the sword instead of Point Four and the jeep.

The scooter could not be repaired, as the Persian mechanics at the motor pool could not make the magical rings and the piston. The Vespa agency in Teheran, which I phoned for the parts, did not have them either, but the agency cabled Genoa to airmail them. In any event, I would have to hole-up and wait.

Major Ritchie then offered me a solution. "Join the boys trucking to Teheran on Saturday. We'll lug your three-wheeler in one of the empty trucks. But it is a long haul, I warn you."

It was, taking twenty-two hours. With most of Iran bare and barren, a wasteland that cruelly identifies itself with death and indifference, I felt how deep indeed is the Persian poet's reaction to the land when he talks of wine, opium, houris, gardens and groves.

It is a static world, otherwise, with Iran corroding in the blazing sun. The peasant, with his mule and camel, has no secret way of life, even though he appears to look mystically at everything. In fact, he is "teched," conscious that he is peculiarly a Persian and therefore different, for he is the humbled man, with nature giving him little and taking away more in the sun-baked stony world he lives in. It is either too hot or too cold. He is a beast of burden and a burden to himself; and so the poets, from Saadi and Ferdowsi to Omar Khayyam, offer him wine, women and opium as a solace. For what else can solace him in the overlit, barren plains and mountains of Moslem Iran?

The Americans were also "teched" in their military gardens, busily escaping from the American blueprints for living. They were the strange foreigners, living with some circumspect fruition, preferring the company of the local Armenians, whose forbears were invited to Persia by the great Shah Abbes in the seventeenth century. The 5,000 Christian Armenians who crossed the nearby border with their arts and crafts had multiplied into one hundred thousand craftsmen and businessmen. They came to settle and live; but the visiting militarized Americans had come for a year or two to see Persia's world dance before their brotherly eyes. They were the soldiers of facts rather than fancies, but they took to fancy in Persia. They bought rugs and tiles, brass and stone, carved gods as well as real goddesses, discovering what Americans always discover with dispatch—that sex away from the U.S.A. is always more pertinent and more exciting; and so they busied themselves with the Armenian girls.

The night before I left Tabriz, two Persian intellectuals asked me over for a drink. Dr. A. worked for the United States Army and was called in for consultations when strange medical things overtook the sergeants and officers. Dr. A. also ran the local clinic, where stranger things overtook the children suffering from glaucoma and fevers, as well as malnutrition and social barbarism. Dr. A.'s friend, Dr. B., had been a student at Columbia University for five years; but his talents were lost, for he was a translator doing liaison work for the U.S. Army technicians in their dealings with their Persian hosts. Dr. B. loved literature, and military manuals not at all, which he occasionally had to translate, and by the rigid numbers. He talked of John Steinbeck as a social force and a literary genius; about T.S. Eliot; and Supreme Court Justice Douglas, with whom he had gone on several mountain-climbing expeditions.

They lived in a secretive world, nevertheless, seldom speaking what they really felt about the Iranian government's fumbling ways. Was it venal and corrupt? It was that, too. It was still opium-sodden, for 75 per cent of the population took to the poppy despite its illegality; and with misery so rampant, its ramifications were all-consuming.

I could have used opium myself to deaden the long drive to Teheran; for the two drivers and I, sweating in the crowded cabin of the two-and-a-half-ton Army truck, tried everything to stay awake in the blistering heat. When our four gallons of coffee gave out, we bought Turkish coffee at Zanjan, pausing to examine, as passing tourists, the Sultanayeh Mosque that makes Zanjan more than a hurried vision seen in flight. It stood, as all things do in Persia, in a self-contained mysticism, which is hardly what it was intended for when it was built as a Moslem edifice for worship. But we were falling asleep on our feet, and with irreligious fatigue. When we reached Qazin, to have a look at the tomb of Sultan Hussein, it was bypassed by the driver. He merely waved an indifferent arm, saying, "It'll be here when I come back, Mister Rooter Scooter," and off he went, rushing through Qazin, ninety-six miles from Teheran. We reached Teheran early in the afternoon. My scooter was delivered to the Vespa agency; and I, suddenly adrift, was looking for a hotel and a much needed sleep.

But it was at a pension, run by a Russian woman, where I slept that long drawn, endless night. Mme. Pars spoke French,

German, Farsi and English; and whenever she passed my hot room, she suggested cold tea or Coca Cola; but seeing me struggling with sleep in the airless room she had deposited me in hours earlier, she made me haul my bedding up to the roof. There I was rearranged and bedded down, in a world of clear air and stars, to await the morning and news about my scooter.

In the morning I rediscovered Teheran or Mme. Pars' pension, both running simultaneously into my vision. I found acrobats, jugglers, tight-rope artists, Chinese cooks turned tumblers, all rushing about her kitchen to make their breakfast. The pension was a refuge for foreign vaudeville performers who worked at night in the dives around Teheran and called themselves "artists" during the day. There was Chen, the Chinese from Shanghai, who had lived in Germany so that he was letter perfect in German, imperfect in English, and compromised by an American; for his former tumbling partner, a German girl, had tumbled off with an American sergeant, who also gave Chen a punch in the nose during the discussion that followed this bit of seductive acrobatics. Chen, who lived in a confused ingathering of junk at the far end of the big flat, was wondering loudly how soon the German girl would tire of Coca Cola and come back to tumbling with him.

Near my room lived two Germans, a lady team of highwire artists; the mother looking like a weight-lifter as she paraded around in her bikini, exhibiting freely what Gypsy Rose Lee exhibited for cash, though her very pretty daughter did not bother to wear anything at all, going about nude at all hours of the day. When she passed my door, I usually stopped typing to stare at her lovely nudity. Once I asked her in. She said she would join me; that she was going off to dress. When she returned she was wearing nothing but a black bra, which most likely suggested modesty after a fashion. The daughter's real act, however, was balancing on her right pinky on her mother's head, neatly balanced everywhere *über alles*.

Mme. Pars had discovered the American Army and Coca Cola at the same time. She rented rooms to sergeants who crowded in every month for supplies; and the sergeants, helped on by Mme. Pars, discovered pretty little Persian girls, the rented wives let out on the installment plan. To some Americans the *sigheh* was a fleshly godsend, but to the Moslem it was a religious concept, after a Moslem fashion, being socially acceptable though frowned on by the Western-minded Iranians who objected to the cultural lag inherent in this polygamous throwback.

It was Mme. Pars, of course, who instigated the train of thought that led me to consider the *sigheh*. She was in the garden, cutting red roses, gossiping with her Armenian neighbor. Did the local rug merchant have four or eleven wives? It was hard to make out the number, but even from my window I could see Mme. Pars' eyes sparkling softly, much like the young Persian girl to whom I had given a bar of chocolate yesterday. She was dressed in a cool black shawl or *chador*. and she nodded good-by as she watched me, a dusty, romantic American, careen off again with the military convoy.

Later on, while I was resting, Persian black eyes looked down on me again as Mme. Pars brought me tea. Her eyes were mocking, sensual and suffering, combining all that is feminine in Iran.

"You need," Mme. Pars said softly, her eyes mocking me more than ever, "a temporary wife, a *sigheh*."

"A wife?" I echoed, pretending that I had one. "What do I need a temporary wife for?"

"She can make a lonely man happy for a little money and for as long as he wants her. You must be lonely."

"But I've only been in Teheran a few nights."

"That is too long to be without a woman," insisted Mme. Pars. "There is no virtue in a grown man pretending to be virtuous," she said simply as she left.

I had read the Koran, of course, which preaches the taking of temporary wives. But I was not sure whether the Koran allowed virgins to accept the status of temporary wives. I vaguely remembered a discussion that a Shi'ite Moslem priest had had with an acquaintance named Ali, who thought of taking a *sigheh*.

My friend Ali had insisted that he wanted a virgin for his *sigheh*. The mullah, or priest, objected. A virgin taxes an older man."

"But I am still a vigorous man," protested Ali. "And I keep four wives very happy in accordance with the Koran."

The priest replied unctuously. "The Prophet Mohammed has declared that sexual modesty and faith begin with virginity. When you take an unspoiled virgin as a *sigheh,* only to have pleasure with her, you injure her modesty and her value to other men."

The next day over tea, Mme. Pars announced, "I have arranged to take you this afternoon to see a mullah, who will get you a *sigheh."*

"But I'm not a Moslem," I countered, still thinking that she was making sport with me.

But her semi-Oriental intrigue was serious. She began by telling me that all men owed something to the women of Persia. Though a Moslem woman of the Shia sect could have but one husband, a man was allowed four permanent as well as seven temporary wives. He was only proving how good a Shia Moslem he was by sharing his bed with eleven women. The fate of Persian women, except those who, like Mme. Pars, were emancipated, was a sorry one. For without husbands or love of any sort, they were eternally single and eternally sad.

There was bitterness in her voice as she explained the role of Persian women. A female in Iran has many burdens. From the plow to the bed she moves completely under masculine control. She is a woman under a shadow, spiritually and physically, wearing a shroud which covers her from head to foot, with only her dark beautiful eyes visible from her purdah garment. She is the donkey laboring in the fields or the servant in her husband's busy house when she is not one of his handmaidens in sexual dalliance. And for all of her married roles, whether temporary or permanent, she has a contract that is as commercial as a bill of sale.

"Does the *sigheh* do away with prostitution?" I asked Mme. Pars.

"What do you think?" she answered knowingly. "The men who go to prostitutes are not boys who want sexual experience, but married men, some of them with four wives and many *sighehs*. Come, let us go to see *sighehs* and the mullah."

We went by droshky, the old-fashioned Russian horse and carriage that contributes to the varied transport of Teheran. We hobbled over cobblestones, past the rushing *jubes* flowing with water, around which men, women and children were making their toilet or washing clothes and vegetables. This was the water trench for everything and everybody. Our horse drank from it as we paused for traffic, with a nearby peasant making an angry remark to our driver.

As we reached the mud-brown mosque, Mme. Pars said, "Please wait. I must go ahead to the *masjid*, to the little room off the tomb room. An agent is first arranging it with the mullah. Please wait patiently."

I waited impatiently, my senses telling me that soon I would have a *sigheh* after a brief *muta* ceremony. *Muta* simply meant "an enjoyment ceremony," which was frank enough for all concerned. I would intone words like *zuwwujtoku, ankuhooku, muttvatoku,* meaning, "I have married thee temporarily." And so I waited, wondering if this seductive session that Mme. Pars was arranging would keep me forever in Iran, my Vespa scooter rusting to the color of an ancient camel.

Mme. Pars finally emerged from the mosque and led me by the hand. I took off my shoes and entered on borrowed carpet slippers, my body tingling with mystery and sensuality. I was about to meet my future *sigheh*.

At the northeast corner of the tomb room, I was ushered into a small dark room. Two wraithlike creatures stood before me, their *chadors* covering them like twin figures from a Witches' Sabbath. The mullah, gowned in black, with a turban and a heavy beard, bowed to us as he entered from the prayer room of the mosque. Which of the two wraiths, who were still nothing but silhouettes in black, was to be my contracted seducer?

Perhaps I was fortunate, for temporary wiving was less

common today than during the reign of the Prophet. Then the *sigheh* rode along with the conquering Moslem soldiers, providing the fighters for Mohammed and the Koran with legalized, religious prostitution. When the Shi'ites and the Sunn'ite Moslems split into two religious factions, the Arabian Sunn'ites abandoned the practice of taking temporary wives, whereas the Persian Shi'ites continued to bed with them in blessed acceptance and honor.

The Second Caliph, Umar, who himself had many *sighehs*, once said of the *muta* arrangement: "The believer is only perfect when he has experienced a *muta*." The mullah reminded me of this. What was good for the Prophet Mohammed and the Moslem saints was good for all men, Mme. Pars accurately translated. The mullah, who seemed quite concerned about my lonely state, talked on, and Mme. Pars continued to translate for me: "Women get almost no schooling in Iran, for psychologically, spiritually and biologically, women need husbands, not schooling. All childhood is merely a preparation for love. When a man signs a contract with a *sigheh*, he promises the girl a *mehiyeh*, or payment for honorable services rendered. So much a week or so much a month or so much a year, and it can be anything from a thousand tumans to five thousand."

I was impatient to see which of the two shadowy creatures it would be; but the mullah kept intoning about the virtues, not the vices, of the *muta* arrangement, as if each word made their mysterious physical charms mount in value.

"Are both of them virgins?" I whispered to Mme. Pars unashamedly.

"I will find out," answered Mme. Pars. There was a series of sighs and titters from all concerned.

"One is and one is not," shrugged Mme. Pars, once again translating some liquid facts. "The prettier one is not, of course."

"How do you know which one is the pretty one?" I asked. "I can't see their faces."

"The non-virgin is very pretty," insisted Mme. Pars. "She

wants two hundred and forty tumans or sixty dollars a month, her agent says."

The two girls now stepped up and the veils came away. The virgin was pale, frightened, and scarcely looked fourteen. Yet her lips were full, her eyes questioning, her body soft, round and languid.

The other, the non-virgin, moved up for inspection. She was very assured, her physical charms too evident. She was taller and more buxom, her features bold and animated; but her eyes were strangely lit, as if she knew something about opium smoking. Perfume radiated from her body.

"For the right man there is no guessing," I said. "How much does the agent want for renting her out?"

"About twenty-five dollars, but one must bargain for a while."

"I must think it over," I answered. "Can you let the agent know tomorrow?"

"For love there is much time," said Mme. Pars, translating back.

In the interim a well-dressed businessman and another priest entered our room. Apparently, unlike myself, they had come to a decision; for they were soon joined by a veiled girl and a second agent. Before I knew it I was attending a *muta* ceremony of a wealthy rug merchant, soon off to Kerman to buy rugs. He was over fifty, and wanted a traveling and bed companion. The girl was very much in her teens.

It was strange, as if I were attending a rehearsal for my own wedding, which might take place tomorrow.

The second priest intoned as the couple linked little fingers, which pronounced them temporary mates, "There is no moving thing on earth but in God is its nourishment." The religious rite was simple, but the civil contract took more time, with the signatures of the girl and the man being witnessed by the agent and the priest.

With a final blessing, the priest said that Islam is against celibacy and considers the sexual abstinence of the unmarried against all human nature. But no matter how many wives a man takes for himself, he must treat them all decently and make love to all of them regularly; for the Koran admonishes every male Moslem in Chapter Four: "And ye will not have it at all in your power to treat your wives differently; but yield not wholly to disinclination, so that ye leave one of them as it were in suspense."

The girl, who came from a poor village in Fars province, wore black trousers beneath a brightly printed skirt. The man was a Teherani and was dressed in a Western-style gray summer suit. Outside the mosque, waiting for the marriage festivities to begin, were three waiters balancing copper trays heavily laden with figs, nuts, melons, *nuun*—the dark thin bread—as well as plates of shish-kebab, cucumbers and bottles of vodka.

I was invited to join the "married" couple at their wedding party. With the waiters trotting on behind, we got into droshkies and off we went to a hotel, where we ate and listened to hired musicians blow exotic music from their wooden pipes. The couple did a *horon,* the same dance I had seen Turkish peasants do along the Black Sea. The girl, who wore slightly misfit shoes—believing this would ward off evil—did a solo, sinuous dance, her hips moving like a houri's. This was her initiation to love, literally, for her virginity showed itself by her moments of excitement followed by sheer physical fright.

"Your initiation is over, Mr. Roskolenko," said Mme. Pars, as we left for home. "Should this girl have a baby, then her husband pays for the child and his 'wife's' upkeep until the baby is seven years old; after that the husband takes the child over permanently. Of course, he may stay with this girl even after he comes back from buying rugs in Kerman. But if he lets her contract run out, she must wait one hundred days before she can 'marry' again, the one hundred days assuring her that she is not going to have a baby from the rug merchant— for it might embarrass her new husband."

"I am not joining little fingers," I told Mme. Pars the next day. "For all I know I might get entangled with American laws back home. Economically, it's a very sound arrangement; spiritually, it's a bit tough on the girl; physically, it's almost like a call-girl setup." And I went on to state various objections, ending up by still pretending that my wife back home (though I was no longer married) would promptly sue me for divorce because I was both an adulterer and a bigamist, which was apparently too much for a quiet American in Persia.

Mme. Pars laughed, then said slyly, "You have too many Western morals in the wrong country. But everything changes in Iran, except that which goes on forever between men and women. Take the *sigheh* you saw married yesterday. She is a poor girl from a poor village. She is no more than sixteen, but many of them get married at thirteen by lying about their age. It is usually their fathers who force them into marriage for the money, which they take, by the way. Before the Prophet Mohammed came, girl babies were buried alive by their fathers; that is, only one of every four girl babies was allowed to live. But the Prophet, who must have been oversexed and liked *sighehs,* taught the Moslem fathers not to kill off their female offspring; and so women believe in Mohammed, for he brought them life as well as legal and religious lovers and some cash. Today a *sigheh* can at least have a temporary husband. Also, Mr. Roskolenko, if you are very nervous, you do not even need the services of a mullah to perform the finger ceremony. All you need do is to face east with the girl, toward Mecca, and say, " 'I myself marry you to myself for a known time and a known sum of money.' At the end of the contract you can renew it if you choose to do so, or you can let the girl go free. You can have her for a day or even one night."

As a final gesture, Mme. Pars, as if to remove her suggestion, said graciously, *"Insallah—*if God wills," and *"Masallah—* God guard you"

Teheran was still simmering hotly. After the denationalizing, the redivision of Abadan and the imprisoning of Mossadegh, the oil burning upon the Persian political waters had gone out. Now it was a city in transition, with many foreign firms basking in the noonday sun, erecting the new commerce, replacing the old traders who once used camels and were now using diesel trucks in their semi-roadless world. Point Four, America's contribution, was busy with the wheat and the nut culture, promising to take care of the needed roads another day. But everything was being taken care of, as Teheran, with jet-propelled speed, headed for the New World in being. And with the Shah visiting Moscow, where he made a significant pro-Western speech as he looked into the Russian bear's mouth, everything was definitely all right. No one had to worry about the Shah turning into a neutralist, like King Sihanouk of Cambodia had done after playing the role of supreme anticommunism in French Indochina. The Shah had honor; he was popular; but the people and the peasants had nothing.

I had my scooter, however, and eight days later it was allegedly repaired. Two rings and a magical piston had been taken from another three-wheeler and substituted, since Genoa had still not airmailed new ones for me. And with the substitutions activating me, I decided to get moving. I left on Monday, the sixteenth of July, heading for Isfahan, via Qum. The road running across a section of the Teheran desert was scaldingly hot. I wore my homemade havelock, scootering like a modern-day buccaneer as I went from mirage to mirage, which appeared when I reached Shahr Rai, less than twenty miles from Teheran.

The scooter was not working too well, hardly making the hills in third, and when I finally reached Aliabad, I knew that I was in for trouble. Oppressed by the heat, I drank almost a gallon of water, but the heat dried my mouth the moment I stopped drinking.

The next stop was Qum, seventy steaming miles across the lonely desert road—and then it happened. The accelerator cable snapped. I stopped a bus as well as a car with four students; and soon the collective mechanical consciousness of Persia was at work. For an hour they messed around with the broken cable; but seeing how vain indeed was their very warm and pleasant comradeship, I took over the job.

"You do not have a woman with you?" asked one of the

students. "That is not right, Mister American. You must always travel with a woman. For instance, right now she could comfort you here. When you go back to Teheran, I will get you a nice fat student to go along with you."

"A *sigheh,* for instance?" I asked, managing to make the cable do for the trip back to Teheran.

"No, a very modern girl, Mister American. She is a classmate of mine, but I think she will have a baby in six months. It would be right for her to have a man; you know, to make her name better."

"Your baby, perhaps?" I asked, amused.

"No, one from a good friend. Would you like to meet her, sir?"

"I'll think it over," I said.

"We will escort you back, Mister American," said the student. "There may be bandits and you have no gun, no? They may steal your camera and your typewriter. It is very nice, that typewriter."

Off I went, trailing them. I lagged behind at half speed and finally lost them and the mysterious pregnant girl. Besides, the student was much more interested in my typewriter. Probably he was a poet without a portable machine, and I was one with one . . . but there was a transaction regarding my current typewriter which had increased its peculiar value.

In Paris I had traded in my heavy Royal for a much lighter Olivetti, which was built for export to Denmark, with a keyboard to fit the fairy-tale Danish personality and language. By the time I reached Teheran, the box holding the typewriter began to break up. The Danish high-wire artist at Mme. Pars' pension immediately made passes at it, suggesting that I take her much better Hermes, which she had bought in Turkey; she'd swap her Turkish language machine for my Danish language machine. We did, after much study, with a typewriter mechanic changing the position of at least a dozen keys, thus allowing me to write a simple English instead of complex Turkish.

I was once again back at Mme. Pars', who insisted that I had a fever. Actually, I had a touch of dysentery, and the source of it was the water I had consumed at Aliabad. I went to bed for two days, feeling very weak indeed, though still anxious to get going as soon as the scooter's nervous system was once again readjusted.

I was a self-appointed amateur ambassador on two wheels, wearing a pair of dirty khakis instead of grey flannels. I became an amateur handshaker, without political ties. For I was in a peculiar kind of politics. I laughed at my own country when I thought and felt that laughter evened up the score, especially when one of our usual diplomatic fiascoes boiled over. A joke at my expense, with a Joe Miller or a Henry Miller touch, glossed over our "national innocence." One day, at the main post office in Teheran, a clerk said to me, "You're always smiling and happy. Are you really an American?" Oddly, I was not especially aware that I was either happy or smiling. Perhaps the scooter had given me a permanent sense of humor, or I wouldn't have gotten this far.

"Are all Americans like you?" he asked later, after inviting me to meet some men and women interested in America, who haunted the United States Information Service library for material they considered to be especially American.

"Are you a happy people, as William Saroyan says poor Americans are?" asked the clerk, point-blank.

"You'd better ask the State Department; it's their secret weapon," I said, trying to laugh that off. In any event I was happy, according to the clerk; but I was a self-styled bachelor on a Baedeker, exploring the insides of a strange map—myself. But the man outside of me reacted with all the psychological American reflexes, though I was hardly the tourist, real or imagined. I was traveling *F.O.B. Roskolenko,* occasionallly making contact with my fellow-Americans as I swamped over into the decorous terrain of the American Army's amiable proving grounds.

The militarized Americans in Iran lived on the largesse of our exportable abundance, their extraordinary amenities giving them a modern, American-aided view of the ancient Persian paradise. If they were married, they had excellent quarters, with a maid to tidy up their littered-up heaven. But there was an enoromus emptiness in their sky. They were, further, suspicious angels in khaki, their energy and zest activating the typical suspicions of the American abroad. He was romantic and cynical; as if the American heritage demanded everything in the world, and after getting it, on his own terms, his world still managed to remain empty. He was, as he put it in his suspicious frames of reference, always "overcharged" and "took," psychologically, culturally, and financially.

It was a useful complaint, giving our ambassador in khaki an endless talking point against boredom. He was the grown-up boy, the product of past Yankee shrewdness and the dollar; but though the dollar did some of his work, his shrewdness lay dormant. Despite his energy, he was lazy mentally, using the dollar's unique value as a diplomatic substitute, which it often is not. The American blueprint that so much of the totalitarian and free world wanted was gadget-ridden, making the khaki-gadgeteer no longer free or carefree as he walked aimlessly about with his low-hanging mechanical ecstasy, just one rung below Nirvana.

It was an imposing Point Four world, the rules governing the nut agriculture we were teaching hardly allowing for wisdom in cultural matters, especially since we ignored most of it, which made us slightly ridiculous in this best of all heavens which we had engineered in Congress. A few daring souls ventured afield, hoping to establish human rapport and with American gadgets as a burnt offering; for their own spiritual needs demanded contact on other levels of life. But strangeness is endlessly frustrating; for there was the language barrier, then the human barrier—the "differences" that isolated the American after he had offered up his enormous Point Four candy bars, the baksheesh the strangers endlessly expected from the Americans, who appeared to have nothing but materialism to give away. For charity is deadly and self-devouring. But if you give all of yourself, healthily, you are accepted on human terms, and I wanted to be very human in Iran. For life in Teheran's foreign establishments was certainly not highbrow; it was no brow at all.

Being human also meant waiting for the reconstruction of the scooter's internal parts. I soon developed the art of taxiing cheaply to the N.C.O. Club on Taktid Jamschid, where I ate hamburgers or roast beef for dinner; for the smell of lamb and shish-kebab was making me violently ill. It was my human reaction to the Persian's perpetual lamb-on-the-spit cuisine, which I smelled all over Teheran and Iran, in every city and village. It came on its magic-less carpet, woven into the atmosphere, stealing over chinar trees and olive groves. It came with dervishes and tea-boys, in strange nuances of people's odors, pungently air-lifted so that it embraced me on every level of activity. When I finally vomited, I recognized this physical strangeness, the violent reactions on my stomach, nostrils, mind and matter, in one fell negation—and so it was temporarily comforting to escape from the over-all lamb diet as I languished at the N.C.O. Club, where I saw soldiers put ketchup into their coffee when they were not putting it on top of vanilla ice cream sodas. However, every country has its fatal failings; and we had an abundance of ketchup in our souls.

There was a lot of comfortable fat around the middle in the Teheran military establishment. Life was a peculiar picnic of work until 1:30 P.M.; then the soldiers went to the N.C.O. swimming pool. It was so routinized, from movies to meals, that I wondered how the more imaginative men under arms stood it. But fat grows easy and the stories grow with the fat. I heard one over iced tea which went like this: Six years ago, during the early American days in Iran, an American general, who had been going with a Persian lady, and in a most soldierly fashion, was suddenly hailed into court and forced to pay the "lady" fifteen hundred dollars. She accused him in court of devirginizing her, though she had been well known in Iranian circles as a prostitute for at least ten years.

Everybody talked about Iranian graft, as if graft had been invented in Persia. Since I'd lived in China, I knew that it was also invented in China and as my home is in New York, it was also invented there. Everybody has his own patent on this rather universal religion. In Teheran, said one irate soldier, "I had to bribe the official notary to give me a permit so I could marry an Armenian girl. It cost me ten bucks." And there was the funny one about Iranian drivers intentionally hugging close to Americans driving military or private vehicles; if they collided they collected immediately, and the sum was always double the costs of the repairs.

"The country has not made the progress of Turkey," said an Armenian girl at the swimming pool. "The Turks had Kemal Atatürk and they moved a hundred years in fifteen. Here the Iranians are, with the exception of Teheran, still in antiquity."

Once more I launched my repaired scooter, happy to be adrift again. I had a companion now, a young Iranian student-engineer on his summer vacation. Hussein was the brother of Hasan, the Vespa mechanic. Over our mechanical parlays at the shop, he suggested that Hussein would be a fine fellow-traveler to have along, for he was soon to be a mechanical engineer. I agreed, happily, and soon I was invited over to dinner to meet Hussein and their two sisters. Hussein was shy, but the sisters were at the staring stage, their lovely dark eyes studying my blue ones. We sat in the cool garden, where the blue tiling accented the evening shades that suddenly enveloped us. The girls, summoning up courage, asked about American girls and how much freedom they had. I said that they were very free, having more freedom than the American man; that we were experimenting with a modern form of matriarchy, since the men died before their wives; and when the laughter subsided, they asked about sexual freedom, much like students at a symposium. The brothers were aghast at the questions, as if this were the first time they had heard their sisters show concern with this problem.

The pool in the garden looked inviting and I suggested that we take a swim before going to sleep, but Hasan, who was the acting head of the family during his parents' absence, demurred. Hussein was in favor, but after sampling the water, he said, "This is very dirty. Tomorrow we will swim in a natural *jube* near the Khuran desert, on the way to Meshed. It will be cooler going by Meshed, though the Khuran desert is very close, as well as the Kavir desert. We shall scooter to Meshed, which is not far from Afghanistan, then we shall turn south toward Zehedan, where we can go across to Pakistan."

We left at 6:00 A.M. to escape the heat of the morning sun. I drove until noon, when Hussein took over, his hands nervously fingering the gears for a few minutes before he started off.

"I have never driven a scooter with three wheels before, only one with two wheels." A few minutes later he almost ran into a truck going the other way. When another vehicle approached, he ran the scooter into the desert, his hands shaking with fear.

"The dust, Harry, I can't see when the trucks approach," he said getting all the more nervous. Fortunately, nothing too serious happened, though he broke the headlight with his antics. When he repeated his desert dash a second time, I took over to keep from getting us both killed.

At Mamazon, about thirty miles from Teheran, I came across a familiar do-good organization, the Near East Foundation. I recalled dropping pennies in their boxes when I was a child, when it was called the Near East Relief, helping victims of the atrocities committed during the First World War. Seeing their sign on the road, I detoured to buy some much needed reserve gasoline.

I met the educational director, Bill Householder, from Colorado, who had been at the Foundation for four years but was going home shortly to take a doctorate at Cornell. I sat in on a class studying the use of dung and modern fertilizers, the more modern stuff being suspect by the young peasants. They preferred dung, probably thinking it had better social and spiritual values, as well as political ones.

The Foundation, which was partly financed by the United

States government, taught sanitation, rural development, husbandry, malaria control, as well as the usual agricultural courses intended for the student-peasant going back to teach in his village. Bill proudly gave me samples of tomatoes and carrots, as well as their own tinned meats, and also three gallons of gas, which he insisted was on the house. Remembering my little contributions to the old parent organization, the Near East Relief, I told Bill that most likely I was getting my money back, with interest.

It was always strange meeting mission-laden Americans off the beaten path. They were doing something for the spirit, and I don't necessarily mean missionary work, though they were men with missions. Bill was one of them, as well as Wilson, whom I met briefly and who was murdered by Iranian bandits a year later.

When an American leaves home he is a socio-romantic seeking his lost frontier, which cannot be found in the American ranch-type communities. Missing something vague or metaphysical, he goes out, as the last romantic, to discover his own White Whale and usually discovers himself in the process.

But the Persian desert was far from romantic. It was brutal, the oven-like gusts of heat throwing up super-heated Dantean waves that almost knocked me off the scooter. I wanted to sleep, for the heat lulled me; but when I saw a jeep and a bus mangled in head-on shapelessness, I stayed awake. Despite the infrequency of the traffic, how had they managed to collide? We shuddered. Hussein made a Muslim prayer, insisting that I join him. Both of us got down on our knees, facing the east. I followed Hussein's little prayer asking protection from bad drivers, from the sun and from the wild animals; and I threw in a request that I find a benefactor willing to take my American Express checks and give me Persian rials in exchange.

By nightfall we reached Samnan, where we tried to fix the broken headlight. But the young garage mechanic we went to played a game called, Now-you-have-a-light, now-you-don't; and it ended, two hours later, with don't.

We found a typical, sprawling hotel, with many beds in

the garden and the usual unusable pool. The owner wanted eight tumans for a bed without breakfast. I promptly gave him five and he was more than satisfied. If you don't bargain, you are a scoundrel depriving a Persian of his national sport. You are also a bit of a fool; so you bargain mightily, if only to increase your social standing.

At this garlanded hotel, I met an English-speaking Persian named Hoda, who worked for the Iranian Oil Company. He was on a mission in Samnan, doing a physical inventory of the equipment left behind by the recently dispossessed Russians, who had finally given up a prospecting oil lease on the Khuran desert which they had inherited from the czarist regime. They had not found oil nor had they been seeking oil during the last fifteen years.

"They had a receiving and a sending wireless set," said Mr. Hoda. "I think they were in another business. In ten days I will have a breakdown of their two oil rigs, which have no value at all as they are ancient copies of American rigs. They left a lot of books behind, most of them technical books from America. Perhaps they were students when they were not being spies? But not one Russian classic, so what do you think?"

Hoda thought a lot, especially about American literature. He also loved Steinbeck, who has a great vogue in Iran, with doctors, dentists, and engineers translating Steinbeck during their off hours. Steinbeck offered them a sort of moral challenge, as they identified Steinbeck with Persia's human wasteland, and Hoda was the third man I'd met making a translation of the same books.

When I asked Hoda about Mossadegh's crying antics during the anti-British period, Hoda grinned and explained, giving me an example of Iranian emotions as he warmed up and gesticulated. "We understand tears and hysteria. It is a national custom to cry when you want to make an appeal. But you thought that Mossadegh was crazy. He probably was, by Western standards; but he was a real Persian when he acted up, winning a lot of sympathy during the nationalization period. It was not his tears but his policy that was wrong. If Steinbeck could cry, he'd be the Premier, if not the literary Shah, with the Jamea Mosque of Samnan renamed after him," said Hoda as we had a vodka nightcap and said goodnight.

The next day was weird, full of low finance and high jinks, with the holy of holies for an American traveler, the American Express check, receiving no recognition at the Bank Melli in Samnan. The bank manager refused to change a ten-dollar American Express check for me, and he also turned down Hussein. This stage of the financial drama was followed up by the manager's suggestion that we scoot back to Teheran and change the money there. It was Kafkan in mystery, ridiculous in its lack of sympathy, with a situation fitting the Persian sense of the mystic. I could only believe that the manager thought we were bank robbers out to club him on the head with our express checks.

My bewilderment gave way to a pleasant discovery, for I found two single American dollar bills in my wallet. Feeling rich, I called the manager of the bank three kinds of a goat and a pig, and I went to a gas station operator willing to gamble on the two "unknown" greenbacks. I got gas and some rials and he got the two bills, which he showed to his friends. They studied the bills as if they were works of art. One even smelled and tasted a bill, probably thinking it a Point Four remedy for hunger, dirt and poverty.

One hundred and eight miles later in Shahrud we spent the last of the rials. Remembering Mr. Hoda, the engineer, I put through a phone call to Samnan and explained, to his high glee, the state of our unique finances; we could either starve, go native, take to the bush or become mullahs, but we could not cash any of our checks. Mr. Hoda, who had splendid reactions, and was a man worthy of translating Steinbeck, offered to bail us out. How much did we want? He would send us a hundred dollars in rials—and not to fret or worry about returning it. I said that ten dollars in rials would get us to Meshed, and that I would send him my check for the amount. A few hours later, after many circumlocutions involving the head clerk of the phone company, and a friend of Mr. Hoda's who lived in Shahrud, I got the ten dollars in rials, which would last us until Meshed providing we didn't eat much and just used the rials to buy gas.

Refinanced, we now took to seeing Shahrud. It was a filthy, noisy, sickening town. The population lived in the streets along with their animals, roaming up and down the main street, spitting seeds before the Tomb of Bayazid, when not urinating before the Tower of Ghazan. And when the manager of the hotel said, "Lock up everything; people steal everything here," I solemnly agreed, especially when he gave me a lock that did not lock and a key that did not key.

Shahrud was not the garden of the Gods; and the houris were not evident at all. The town was filled with blind men and beggars who appeared to be its major industry. I remembered what a friend had said in Teheran, when I was praising the amount of new industrial projects going up: "Wait until you see the rest of Iran." How right he was; for deserts do not bloom in Persia. Men rotted away in the towns, corroded to extinction by indifference and heat.

We fell asleep, finally, adrift in this "teched" town with the radios whining and lamenting from every café. And through the night I kept seeing Hussein running into trucks; then Hussein sick—then I, much sicker.

In the morning, I was handed a note by the manager of the hotel. It had been written by Hussein and said, phonetically and dramatically, "Mr. Harry—I go back to Teheran because I sek and have not money geste two travel cheques. I think Bank Melli in Meshed no cash these cheques. Hussein."

The vibration on the corrugated roads had given him fever, and I had not realized how sick he really was. The night before he had refused to eat or drink. He was odd, too. Once, when asking a truck driver where a gas station was, and being told it was only a mile away, Hussein had stormed, saying angrily, "That man is a liar. He is crazy and corrupt like all Persians. I think I am crazy too." I had heard this so often that I ignored Hussein's self-indictment. He was not crazy, nor was he corrupt, though Iran had more than enough Persian experts in that field.

Currently, corruption was getting some prize comments in the Congress of the United States regarding a recent visit made by an investigation team of congressmen who came to Iran to see what had been done with a hundred million dollars given to Iran by the United States.

"We built roads with the money," the congressmen were told.

"Good, but where are the roads?"

"They must have blown away," said one Teheran columnist, "for we have very big winds in Iran. And the winds come from Russia, which does not help us at all."

They should have asked me, a down-to-earth expert on roads Iranian.

What could poor Hussein do with his Bank Melli checks? He had checks enough to keep him for two months, but two days after leaving Teheran he was literally a beggar, like most of Iran's population. I read the note over several times and wondered what I could do for him. He had taken a bus at 7:00 A.M. for Samnan. I tried phoning the Vespa agency in Teheran where his brother, Hasan, worked.

"There is no telephone," said the man who ran the radio telephone in Shahrud. I insisted, saying that I had phoned the agency several times, but my insistence did not get a connection to Teheran.

Should I go on alone? Soon I would be entering the bad, mad lands, warned not to travel there alone or at night. I decided to chance it and do little night riding. I packed, ready to leave, only to remember that the headlight had not been repaired; that the bottom was falling out of the "station wagon attachment." It was hardly a station wagon, but it was an attachment: a few boards nailed together, making a small cabinet or box to hold my sleeping bag, extra tire and my tools.

After many hours, using three local carpenters, I managed to get the box repaired, knowing that the corrugated roads would soon tear it apart, along with sundry other mechanical parts, including the driver. With some sentimental pangs, I wired the Vespa agency in Teheran offering to trade my three-wheel scooter for one of their two-wheel scooters. A two-wheel scooter would take the corrugations in its stride. Meanwhile, a young Armenian doctor offered to take me to Nishapur in his ambulance. I accepted, especially when he told me that for ten dollars I could get my scooter transported to Meshed, 360 miles away, and Nishapur was but 80 miles from Meshed; also, in Nishapur I would see the tomb of Omar Khayyam. Omar Khayyam settled it. If the agency agreed, I'd ship the scooter back and return to Teheran for their two-wheel one. A few hours later I saw my 400-pound scooter loaded on top of a bus.

We found a spot at the edge of Shahrud, pushed the scooter to the top of a small rise, drove the bus alongside the rise, then lifted the scooter onto the top of the bus. This was hoist engineering with a new twist.

"We will leave in the morning," said Dr. Hovanissian, my Armenian host-to-be. "You look very tired and sick. If you drove that thing to Meshed, you would get very sick."

I was sick, soon after we started. Again dysentery, but much worse than it had been in Teheran. I got very weak. A few hours later I was as feverish as the overworked engine of the ambulance. At Sabzawar I lay at the entrance of a new hospital, which was not yet receiving patients. I drank tea and took all sorts of pills, but I was burning up. I wanted cold water by the gallon, but Dr. Hovanissian would not give me any.

When the engine cooled off, I got up from the ground and staggered back to the ambulance. I fell twice en route and Dr. Hovanissian picked me up. When we were on our way again, I tried to straighten up, asking all sorts of medical questions when I was not stopping the ambulance to go behind a bush. For dysentery doth empty a man of his mind and his body. Soon I learned all about "bush medicine" and his country-doctor style in this Persian outback. He mainly treated road workers. Every kilometer had a worker shoveling gravel from one pile to another pile, for which he got about forty cents a day. The road workers received free medical treatment and were always waving the ambulance down, demanding that Dr. Hovanissian stop and give them pills. They were sick; they were dying; they were pretending; they were boondoggling with aches and pains everywhere. They waved and waved, at least three hundred of them, all the way from Shahrud to Nishapur, but the doctor ignored them.

"If I stop for one I stop for all, for that is the way it must be. Would I ever get back to Nishapur? I am given only three days to make the trip, to treat some really sick patients. These men just want to talk and get some pills. When I ask them what hurts them, soon they are pointing at their heads, their eyes, their hair, even to their shoes. I am not cruel, Mister Harry, but I know these people. Besides, if they are really sick, all they have to do is call for the local ambulance, which will take them to a hospital."

The doctor got four dollars a day on the road. He also ran the hospital at Nishapur, where I was to be his patient unless I got better. He had a staff of six medical aides, who got two dollars a day when they made the circuit from Nishapur to Shahrud. Obviously, the medical profession in at least one country was not overpaid.

It was a gruelling ride and I wondered how the doctor managed it every month. He looked as bad as I did when we finally reached Nishapur about 6:00 P.M. He suggested that we go to the local bath, for I was filthy with dust, mud and dysentery. I managed to wash myself a bit and fall into bed. I was feverish and I talked throughout the fourteen hours that I slept. I ground my teeth, tossed the covers off, and made nightmare nonsense in several languages, the doctor told me the next morning.

It started again. More pills, rice and tea. But my body was flowing away, and for two weeks I lay in the garden on an iron poster bed, evaporating with the heat and dysentery. When the garden got too sunny, I hobbled inside to lie in a cool room.

When there was a respite, I ate; but soon I was burning up again, with the good doctor becoming frantic, fearing that I might die. His pills were for Persians, not for Americans. The Persians had built up an immunity; I had built up nothing but high fever. I must be transported to the American hospital in Meshed, and he would arrange that. But I did not go.

I began to get better. I was terribly thin. Protruding bones, where solid flesh had been, surprised me with the vision of myself. I was, by bone structure, a mesomorph who had quickly turned into an ectomorph. On the tenth day I felt that a few bones, my stomach and my head somehow belonged to me. My joints no longer ached, though I was still running off to the squat-down Persian toilet.

"No water yet," insisted Dr. Hovanissian. "Only tea, Harry." It was always tea.

When I felt well enough, I decided to see the tomb of Omar Khayyam. But first I went with Dr. Hovanissian to his office to see him in action with his patients.

"They suffer mostly from dysentery, just like you, if for other reasons. But in this region we get a lot of Malta fever. The milk is bad here and everybody is very poor. Only last week a young man came in and said, 'Dr. Hovanissian, give me money or I will kill you. I have not eaten even bread for two days. If I kill you and they catch me then I will at least have bread in prison until they hang me'"— and Dr. Hovanissian paused, looking sadly into Persian space, before saying, "I gave him two tumans, which is worth about twenty-five cents. He will eat for two days and then he will try this again. In a few weeks he will be caught. It's very sad in this country. We have all the roads open to communism. In Iran everything begins with corruption and ends the same way."

Dr. Hovanissian, who wants to go to the United States to study American medical practices and methods, said later, "My degree from the Teheran Medical School is now recognized in the United States, but I am terribly ignorant. I want to be a doctor, not a medical aide, so I must go to Johns Hopkins. Do you know anybody who can help me get an interneship? I will start from the beginning, just like a student." I mentioned some Armenians of repute, including John Hagopian, Reuban Mamoulian and William Saroyan, who was an old friend; that I would write to Saroyan, though Saroyan's dream world and style would hardly be an endorsement within the medical profession.

At his office, I saw several small girls, no more than two years old, suffering from dysentery. One of them was remarkably beautiful, but with the saddest eyes I have ever seen, as if she already understood what it was to be a Persian living amid sun, stone and weeds.

Old men came in with swollen abdomens or swollen legs. Others had trachoma, but everybody got injections. The doctor sterilized his needles in a small pot, which he heated over an alcohol burner. His equipment was sadly lacking, with not one gadget in his dusty office. Dust was the major ingredient for medicine and pharmacy, with the winds blowing Iran's deserts and mountains into the villages and the cities.

"When did the Armenians come to Iran?" I asked, wanting more accurate data, during a tea interlude between patients.

"Shah Abbas invited 5,000 skilled Armenians here about 400 years ago, but we have not grown greatly. Like the Jews, we have persecutors. In Nishapur, which has almost 25,000 people, we are only 30 Armenians. But this is a very corrupt country. Why, when the Russians came here at the beginning of the Second World War, they shot everybody who tried to corrupt them. But that was only the beginning. Within a half-year they were like the Persians, taking bribes all the time." Digressing, he went back to his family history, saying angrily, "I had an uncle in Tabriz who was kidnapped by the Russians because he was an anti-communist. For eight years they kept him frozen in Siberia, but when Stalin died they let him come back to hot Persia."

During the long journey from Shahrud to Nishapur, while passing through sixty miles of the waterless, grassless desert near Abbasabad, Dr. Hovanissian had told me about the unfertile land and its people. "They have some Georgians here; you know, just like Stalin—and they are terrible people." When we stopped for gas at Abbasabad, a few of the "terrible people" clamored about the ambulance, trying to sell us pipes, cigaret holders and perfume containers. The objects were decorative and well designed, but the clamor and the salesmanship, as each ancient Georgian outdid the other in abuse, was terrifying. I finally managed to select five objects from the least offensive of the haranguing Georgians and asked Dr. Hovanissian to ask what the price was. Automatically, Dr. Hovanissian offered the entrepreneur twenty rials.

"I know these thieves very well by now. When I came through here last month this same man cheated me by about four hundred percent. Here, you thief!" barked Dr. Hovanissian, pushing the money into the protesting hands.

"These Georgians came to Persia as Christians, but they turned into Moslems because it helped them. They would turn into anything, even Hottentots," continued Dr. Hovanissian.

A few hours later, when we stopped for the good water that came down from the 11,000-foot peaks of the Kuh-i-Binalud and was not yet contaminated inthe *jubes,* we saw a busload of Pakistani pilgrims bound for Mecca. Then men were tall and colorful, hardly drab-looking like the village Persians. They sat beside the bus and smoked their water pipes, looking sagelike and patient, for the trip to Mecca took several weeks by bus. Their women, who wore face coverings, with lace openings for their eyes, were protected like rich baggage and treated as such.

Driving in Iran was obviously a hazardous business. We saw a huge truck over on its side, its cargo of broken glass being shifted to another truck.

"The driver dozed for a moment and this happened," said Dr. Hovanissian. "If you doze, you die."

Feeling much better, I went on short exploration trips, starting with the nearby tomb of Omar Khayyam. It lay in a grove at the end of a garden shining with red roses. Nearby was a serrated plain, with corn and hay waving in the hot winds coming off the Khurasan desert. It was a garden of silences, literally, waiting for the subjective visitor to give the tomb meaning as he voiced his emotional reaction or remembered stray lines of the *Rubaiyat*.

"Khayyam" meant "my eggs" in Farsi, which is hardly what I expected, knowing the fatalistic yet romantic flavor of the *Rubaiyat's* inner concern with the Persian spirit. The earlier poet, Ferdowsi, had been better named, though it took a Sultan to change Mansur Ibn Ahmad's name to Ferdowsi, which means "paradise." There I was, between the eggs but not in paradise, viewing the roses bending in the sultry winds, trying to visualize the past, when Omar Khayyam's lyricism gave a dying country its last literary marvels.

A cobblestone path led directly to the over-simple tomb, which had the poet's name inscribed on the front and some lines, probably ending with "Paradise enow," inscribed on the back of the tomb. The tomb was twenty-five years old, having been built by the Reza Shah to replace the original 800-year-old tomb which had decayed, along with Persia, centuries ago. Near the tomb was a blue and brown seventeenth-century mosque, with the usual "Baksheesh" dignitary standing at the entrance ready to take my shoes and give me a pair of carpet slippers in return. But my religion was closer to lyrical poetry, not Mohammedanism, the silences tempting me to other echoes. I visualized pagan rites and I heard houris singing. I saw the Persian world of 1220 bursting all about me. Here, in the effervescing heat, men and women were dancing to the sun, leaving the sandy, corrosive death to the Khurasan desert. The old poet's symbolism and philosophy gave me an hour of imagined greatness. He was fleeing from the shaft of marble; and I heard the soft, mellowing, burning lines give me back the world of violent roses and wine, where the sun gave life and death at the same time, blossoming in these fields of roses much as if they were natural poems. And Persia was a rose when it did not crucify man with its deserts. It was nature lamenting for all that man was and still is. I recalled my humility when I saw Mount Ararat, but I was looking at the Kuh-i-Binalud, which was less inspiring in stature and meaning. I kissed the tomb and I went away, the sentimental man holding a red rose and some yellowing sand in his hands.

The men in Nishapur are henpecked, especially when they visit Dr. Hovanissian with their womenfolk, who berate them in public but end up getting beaten in private. The doctor's shabby office is always crowded. And when the doctor gets a hurried call to visit a patient too sick to come to his office, he goes to the bamboo-laced window and calls down: "Mahmoud, come and make two rials." Little Mahmoud, aged six, with gay eyes and enormous personality, hurriedly trots barefooted up to the office. He smiles like Roosevelt, perches on the balcony, dangling his feet with reckless bravado and listens to the doctor leave last-minute instructions.

"And remember to report everybody who comes in, even if they don't tell you their names or if they say that they have no name—if you want the two rials," warns Dr. Hovanissian.

Later, when Dr. Hovanissian will ask Mahmoud, "Tell me, who was here while I was gone?" Mahmoud will smile and chirp out, "A woman, who said that her husband did not beat her enough, so she's very sick. Her name was . . ." and Mahmoud will invent a name on the spot, adding, "She also said that she has a hole in her stomach," something Mahmoud has also invented in his passing fancy.

The doctor had his medical problems, particularly when giving injections to his patients. A man recently protested about his wife, who had an injection in the left side of her buttocks for headaches caused by glaucoma, saying, "Since my wife's head hurts she should have the needle put into her head." Injections are magical to Nishapur's many patients and they want them whether they are well or sick; for with an injection you are at one with sun, jazzed up or slowed up, doped up and made momentarily content.

But Nishapur is more than just quaint. Before an earthquake and Genghis Khan ravaged all of Khurasan, Nishapur was the capital of the province, a city going back two thousand years, containing at its height hundreds of thousands of people.

Now its twenty-five thousand ragged population engage in handicrafts, with little boys and girls weaving rugs for twenty rials a day, their voices following the rug master's as he intones the design they are making: "One red; two yellow; three green," with the little hands moving the large bone needle until the design evens up at their section, their small squatting bodies jerking with each word uttered by the rug master. They sit in rows as if they were praying at a mosque, waiting for the master's voice to call to them, or they fill in the simple stretches until their turn comes again.

A twelve by eighteen foot rug takes four boys and a teacher a year to weave, with the price, if it's an ordinary rug, under three hundred and fifty dollars for the year's work. With such pay the children can only afford to eat bread and cheese, which cost them four rials; for meat is something others eat. Dr. Hovanissian soon has them as his patients.

But the rich live well in Nishapur. They grow cotton, have semi-European homes, large cars, water that flows from a tap, and even Western-style toilets, for squatting is hardly a sign of wealth. The brothers Harootunian have such a home. One of them is an engineer schooled in Germany; his German wife, Greta, still feels lost after sixteen years in Iran, where she fled in 1940 with her Iranian-Armenian husband. Now she wonders if her children, who are a mish-mash of nations, will ever have a real nationality or a religious identity.

Her home, in the traditional Iranian style, is walled in. We dined on the large patio, drinking wine and vodka, eating delicacies from the delicatessen along with assorted Armenian-German dishes. The talk was mainly about the nearby Russians, who disturb every moment of the Iranian sense of time and politics; for the Russians are the nightmare neighbours haunting the adjacent mountains and deserts.

The older Harootunian is a big cotton planter, and the younger an engineer who builds factories and homes. In their restricted way they make up the social conscience and the liberal ethics of Nishapur, mixing their Christian values with the Moslem inconsistencies inherent in the province. They are wealthy by any standards, for the house has enormous, carpet-laden rooms, extensive gardens and swimming pools, with grand pianos and paintings giving a European air to the heavily-laden Persian surroundings.

"If we went back to Germany," said Mrs. Harootunian, "we would be poor indeed. But this is such a sad place to live. What does one really have here? They play poker, that's all."

"Monday will be a big day," said Dr. Hovanissian as we were leaving. "The railroad from Shahrud, which is 210 miles away, will open a spur to Nishapur. It will cost twenty-five per cent less to travel. The railroad is coming to Omar Khayyam, Harry." There was a big laugh and another vodka for the road, with Mrs. Harootunian saying, anxiously, "And after the railroad, the Russians."

The railroad rode into ancient Nishapur, and stayed. The people came on donkeys, horses, bicycles and bare feet, to see the first diesel train slowly moving toward the station. The station itself had been built fifteen years earlier, but during the war the Russians used it as an office. When they left, they took the plumbing, the doors and the windows along with them.

But now there were tracks and there was a train outside the station. The visitors staring at the impressive wheels and gears of the locomotive were dressed for the celebration; for more patches, like the fresh paint on the train, add color. The officials and members of the Majlis, the generals and the men who ran Iran, stood on rich Kerman rugs on the platform. The peasants were on the other side of the track, wired off at the far edge, as if contact was unhealthy for all concerned.

It took 2,300 men two years to lay the single track from Shahrud to Nishapur. Engineer Fatemi, in charge of the track laying operations, beamed and said proudly, "It was hand laid all the way—210 miles—without machinery to grade or put down the ties." He made me proud, too, and we toasted to the event with vodka.

Inside the station were tables laden with Persian delicacies.

The celebration was about to begin, for the train was finally coming into the station. The large searchlights glared into the sun, colliding like old and new planets in similar paths of speed and light. Everybody shouted, especially the tattered peasants. The children and their mothers, dressed in the traditional chadors yelled the loudest. The train was the new world, rhythmically fusing with the old, joining the past as it finally echoed explosively into the station, as if welcoming Omar Khayyam himself, who had written:

> With them the seed of wisdom did I sow,
> And with my own hand labour'd it to grow:
> And this was all the harvest that I reap'd—
> 'I came like Water, and like the Wind I go.'

And like the wind, the symbols suddenly changed. The bayoneted police started a harvest of yelling and pushing, for everybody wanted to touch the train. From their wired-off world the peasants could only stare at it, and those who touched were soon sorry, for they were holding their backsides where they were hit by the police. It was better to see just a little, to shout all the more, and then go home. In any event, it would be years before they could buy a third-class ticket or see Teheran, if ever they had the money for either adventure.

It was just another American gift, with General Motors donating it. Not ironically, General Ansari, the minister of roads, spoke in favor of more roads. Dr. Shademan, the exchequer assistant to the Shah and the deputy administrator of the Shrine of the Eighth Imam, in Meshed, spoke about money and those who honored the spirit with their donations. An hour later, Dr. Saadi, a literary critic, made more suitable analogies at the tomb of Omar Khayyam. He spoke of Persian poetry and poverty, as if they had a kinship for each other through psychic phenomena. The railroad had come to Omar Khayyam.

That night, as if haunted by the spectacle of poets, donkeys and railroads, I walked back to the tomb. The railroad was too strange. It had yelled its way into July 23, 1956, but I was still listening to the thirteenth century chanting its ancient accents. I was confused, of course; for I was the typical American who had lived all of his life with gears. I wanted another glimpse of antiquity, and I was transfixed as I walked to the tomb with my sleeping bag, prepared to sleep the night at the foot of Omar Khayyam.

Others may have had the same feeling, for I was not alone along the road. A few young couples, from nearby Meshed, were struggling with their Western-style attempt at being lovers. They held hands, curiously uncertain of their new behavior. It came with the diesel, with Pullmans, with club cars; it had another set of psychological norms.

By the time I reached the tomb, I was talking with a couple, my few words of Farsi and their few words of English enabling us to communicate. They were meeting a girl at the tomb; to pay homage once more before starting by car for Meshed. They were students, lost in their acquired Westernization, lost to their Persian ways, living between two sets of compulsions and cultures.

I was lost, too. When they introduced me in the darkness to the other girl, who had been waiting at the steps of the tomb, as if giving them time to be alone, I felt all the more strange. I was the Western man, however. The girl had been half-Westernized, but she came from the world of the *sighehs,* the temporary wives.

But something was happening, and I was really alone. There was no girl; not yet, anyway. I was within myself, trying to induce a state of fantasy—to see and feel the acute things that had made me the incautious traveler and adventurer, the man purging his conscience and his emotions. And it was a fantasy that had me.

In my fantasy I was walking slowly in the broad daylight, for the Persian summer heat inflamed the road out of Nishapur, bouncing off the Khurasan desert's rocky encrustations. In the distance, the Kuh-i-Binalud soared to 11,000 feet in a cool embrace of high mountains, icy tops and spuming clouds. I could make out the white peaks. The sun glowed there, and the sight of the peaks made me feel much cooler.

It was too late in the day to be out alone; for only last week three Americans had been murdered not too far away. But the tomb was just up ahead, the tall thin slab at the edge of the rose garden, with Omar Khayyam's name engraved in Persian characters. Near the tomb was the old mosque made of brown earth and clay; at the entrance often stood an old man who rented out carpet slippers to the Moslem worshippers. But he was not there now. It was too late even for him; or the mosque was too far from Nishapur for safety.

A bird flew over the mosque and alighted on the tip of the minaret. Awed, I stared up and hurriedly opened my camera. I got the picture in time . . . then I heard voices. Was it the muezzin come to call the next prayer for the faithful? I saw no one. Perhaps the voices came from the white slab, for anything could happen in mystical Iran.

I felt dizzy, for the heat had not let up. My eyes were fogging, and I saw mirages cooking up from the desolation. I reached for my water canteen, drinking long and hurriedly, then continued toward the tomb.

I staggered as I reached it. It must be the fever again, I thought, the bad water I had drunk from the *jube* a few days ago.

Again I heard voices; female voices, then singing.

At the tomb I braced myself against the slab, trying to throw off the feeling that I was about to sink away or fly off into Persian spaces.

The voices were coming at me again, surging and swirling, as if dancers were moving and singing at the same time—houris of the Moslem heaven, cavorting with the blessed, acceptable citizens of Allah.

My Spanish wine flask, my camera, and my sleeping bag had fallen in the grass to the left of the tomb; for in my eagerness to grasp associations, I had intended to sleep there for the night. Sentimentally, I had planned to drink wine and pay homage to Omar Khayyam; to spend one long night in the atmosphere of the *Rubaiyat,* eight hundred and fifty years past the time the poet wrote his ageless, aching lines:

> Whether at Nishapur or Babylon,
> Whether the Cup with sweet or bitter run,
> The Wine of Life keeps oozing drop by drop,
> The Leaves of Life keep falling one by one.

The voices came and went, ringing violently in successive echoes of lament and joy, each mood engaging my fancies, penetrating the weird state of my mind and body. Now I was burning with fever, shaking and trembling; and I sank to the grass. With an effort, I undid the sleeping bag and rolled it open. Then I lay still, my hands curled inwardly, gripping the phantoms of my fantasies.

The voices joined me again; gentle, feminine voices singing soothingly, as if I had ascended the Moslem heaven and found the promises of the Koran about to be fulfilled. Echo upon echo broke through my semi-coma, voices from some twelfth-century world radiant with sensual surfaces and delights. I was in a world of hypnotic graces, and I was as many-handed as the Indian god Siva, yet hardly able to avail myself of all the joy twisting and turning before me.

I heard one voice rise above the others in a liquid, urgent appeal. Now the other voices died away, and I was left with the new intimacy, between religious grace and sensuality, both mingling with each other.

Someone was kissing my lips and my mouth responded. I moved a few inches, and I felt the presence of a woman as she joined me in the sleeping bag.

I heard the Spanish wine flask scrape against the marble, and my mouth was wet with wine; for my mouth was like the Khurasan desert, the heat fevering my throat.

Suddenly I was embraced. I felt lewd and unashamed, holding her to the grass. Then I was adrift again, floating off to the icy Kuh-i-Binalud, to the cool tombs of the mountain tops . . . then my mouth was filling again, the wine was running over me, and I was back again on the hot plains, lying near the tomb.

What else had Omar Khayyam once written?

With them the seed of Wisdom did I sow,
And with mine own hand wrought to make it grow:
And this was all the harvest that I reap'd—
'I came like Water, and like the Wind I go.'

The words repeated, then changed, altering the meaning . . . and my body moved again, crushing into the earth, over the female softness below me.

I live like Water, and like the Wind I blow.

For like the wind I had come to Iran to escape my multiplying American loneliness after my wife had divorced me. Now I saw her, two years later, coming up from the mirages on the Khurasan desert. One single lovely woman. And now I had the heavens populated with wondrous houris, in long lines and successions of the flesh.

But had my wife been solely that? Had she been anything at any time? Had she been the spiritual woman of merit and honor? Was any woman that? Were not American women merely functional machinery, propelled to psychoanalysts' couches by the imagined torments of their all-or-nothing world? What had the good Dr. Freud done to the psyche and the imagination, if not to the loins of the American nation?

The voice began again, subtle and soft, numbering the simple pleasures I had missed during my seven years of marriage. Of course, it was connected to erotic and neurotic symbols; and it had to be those or I wouldn't be the modern American male.

The one single lovely woman—how had she said all things to me, all things that mattered to a man? I recalled, as if it were happening now, that my wife had a breathless belief in psychoanalysis. For one night, over tea, she said, "Harry, you ought to be analyzed. Then you'd begin to earn twenty thousand dollars a year. You have the talents, but you must become aggressive, really aggressive. Be analyzed, then go in for public relations."

On another occasion, she said, just as I was falling asleep after an eighteen-hour stretch on a political survey I was completing, "Harry, every man we know who has been analyzed has a greater earning power today. Harry, are you asleep or are you pretending?"

I was pretending; but then, I was always pretending to her.

The one single unloved man, who had to come daily with good tiding of wealth. And that was America, a new country, but all used up and suffering. Ah, reality! Ah, Excelsior! Ah, newer standards! Ah, everything.

Once my wife had said, "Darling, I'm keeping a separate account in the bank. Darling, I'm using my maiden name on some things—do you mind? Darling, I don't like to make love in the morning, when I'm dreaming. It disturbs my dreams."

The one single lovely woman. But bitchery was the purest symbol; women who took more pleasure in shopping than in the delights and certainties of love, though they knew all the clinical words of the Viennese doctor. They shopped for love in the bargain basements of their emotions, and then went to the doctor's many apprentices for the Talking Magic. And soon they were talking strangely, using terms like "psychic impotence," "frigidity," "id" and "the unconscious." They loved the token words written by a man who had an inventive mind but who could have done better had he written poems for children. But then, Freud was a poet lost in his own vast unconscious.

But Freud had never been to Persia; he had never slept in a rose garden of the mind, but a blue-tiled pool of the body; he had never known the ways of non-Christian-Judaic love, the deep exoticism of the imagination and the body mingling in one great orgy. Dr. Freud had known some Germanic-Judaic-European clinical data, considered to be profound when not dubious; he had known Neurosis, "The result of a conflict between the ego and its id."

I groaned, my memory fevering me all the more. Once my wife had said, "Darling, I think malice is a greater emotion than love; I think hatred moves more people to do better; I think children are for other people, not for me, darling. I love to see my aunt's children, for all they require is a kiss, a pat, a slight embrace," and she had embraced me as if I were one of the children, her hands sweating, deadly and feelingless. It was all undisturbed loveliness, without a sound coming from her lips and nothing from her beautiful thighs. I lay within the surfaces of her mind, upon layers and layers of artifice covering her inwardly and outwardly. I was a stranger living within fearful intimacies, never knowing when her superficial "darling" would some day no longer sound for me. She had treason careering in her soul.

Ah, the good old doctor had it all blueprinted for her. But had Freud known the Persian *sigheh,* the temporary wife? But Freud knew about the Western wife, mistress and whore. He knew *security* and the contagions of emotional dread it inspired in every American.

I, Harry, was like the Hunafa, who had turned from Arabian idolatry, seeking the One God but not the One Couch. For good Moslems the One Couch was hardly a joy forever . . . and I heard houris singing to me again. But hadn't the psychoanalysts put the Prophet Mohammed on their clinical couches, too, in some sort of competitive envy about Mohammed's revelations? Mohammed had said: "Marry of the women who seem good to you, two or three or four; and if ye fear that ye cannot do justice to so many then take only one or the captives that your right hands possess."

Mohammed, at Al-Madinah, had heard voices, and this was revealed to him. What fevers had raged in Mohammed's mind and body, I wondered, trying to stifle the varied associations springing to me. "I am possessed and I possess a thousand religious houris; but perhaps I am only sick and feverish . . . the *jube* water have swallowed me with myself and my fevers."

I finally managed to get up, reaching for my canteen in a staggering movement, so that I fell back again. I tried again and stood up straight, my eyes ranging from the white slab of the tomb to the heavy darkness that had swallowed up the mosque. But I saw no one and the voices were no longer singing to me; not one voice and not a thousand. I was in a silent place, with only the wind turning through the rose garden in front of the mosque.

My canteen was on the ground, but not my Spanish wine flask. The camera was there with the viewer still open, but my sleeping bag was also missing. I searched, dragging my feet around the base of the tomb and found a string of blue Persian beads, the kind given by a friend to protect one on a long journey.

I picked them up and put them around my neck, feeling a sudden dizziness not related to my hours of fever. A perfume came up from the blue beads; a smell I had learned to know well whenever I was with Persian women: the rich flowering of odors that almost overcame me in their presence.

It is very strange, I thought. The dirt track leading to the tomb showed only my shoe marks. Had a thief come while I was asleep and taken my flask and my sleeping bag? What strange tokens! Why not the expensive camera? Why not the water canteen? Why not—and hundreds of flooding images raced through my mind as I left the tomb to find my way back to Nishapur, bulb-lit and sparkling in the darkness.

Had I been out of my mind, fevered into Persian fantasies? Who had done it? Was it the old man at the mosque? But then, the old man did not use pungent perfume, though perhaps he used opium like so many of the Persians, when they wanted more fantasy and more dreams. Had I taken opium without knowing it? Was the water and the wind and the dream a self-induced opiate?

"I haven't smoked opium," I said aloud as I walked. "I have traveled in myself, the American who has everything and nothing."

No, I hadn't smoked opium; but something had happened. I felt purged, cleansed, newer—the impact of the fantasy creating something solid in my mind. The past, which had pushed me from one emotional nightmare to another, was past. I was myself and stronger; for the fantasy had taken me within myself and out again. I had emerged.

The next morning I said good-by to Dr. Hovanissian, having written letters to every Armenian I knew, hoping that this would help him to re-intern medically in the United States. I took the bus to holy Meshed, but my thoughts were far from holy. The girl I had been introduced to in the darkness had my sleeping bag. She had taken it as a token, and I had no intention of asking for it—not after the wonderful all-releasing fantasy that she had, in her own way, engendered. But in Meshed I was soon bogged down by things that shook me differently. I was invited by Dick Snowden, the American vice-consul, to share his quarters until the exchange of the three-wheel scooter for a two-wheel one materialized. Anything or nothing was possible with Persian red tape, which had been subtly adopted from the famous French masters of the medium.

Snowden's clerical assistant, young Frank Forgione, told me that I was crazy; it could never be done. Over a Tom Collins and a swim in their pool, we argued over roads and vehicles, and he summed up the substance by saying, "Dick and I went to Afghanistan a few weeks ago, to see if we could find the missing Peter Winant. Our jeep broke down twice. We wrecked an axle, too; so how in hell do you expect to get out of Iran, which has better roads than Afghanistan, with that thing of yours—no matter whether it is pushed by three or by two wheels? We had four, but it got us up the creek."

"And you also have a matter of the old Iranian Customs, you know, or don't you?" added Snowden. "Anyway, Frank, get going on it or Harry'll be here until the snow comes."

Forgione went to work. A reply telegram from the Vespa agency in Teheran said they would exchange scooter for scooter, but the Iranian Customs were insisting that I get a new export plate. From where? From Genoa, where I had bought the scooter. That would take weeks. I would also need documents from the police, a new carnet, and many things new for many things old. At least ten different government offices would have to be visited, each office nimbly operated by a bureaucrat who specialized in his private brand of confusion. The important thing was to get started; to get the three-wheel scooter back to Teheran, and to fly back—to get going. But what about Peter Winant, who had got going?

The American, Peter Winant, and a blonde Swedish girl were hitchhiking from India to Sweden to get married. They were last seen near Kabul in May, walking with their rucksacks. Winant was related to John Cooper Winant, the former American ambassador to the Court of St. James, and so the disappearance had taken on more importance. Had they gone to Russia? They had no visas. Had they been robbed and murdered? They had little money and a tight schedule; but the girl was blonde and very attractive.

I recalled a few hours in Paris, when a blonde Dutch girl, over too many drinks, argued by the hour that I take her along; and I remembered saying, "It would be like committing suicide; you are too blonde and too beautiful. You have a fatal attraction."

Dick Snowden and Frank Forgione had been unable to find a trace of Winant and the fatal Swedish blonde; it was as if they had disappeared in a traceless world, with the Afghan and American authorities finally convinced that the pair had been robbed, the girl assaulted and both of them murdered and buried.

Snowden, who was a scholar of odd facts, said, "They send Mormon farmers here to help the Iranians with their soil problems. It seems that Utah looks like Iran, in part. It must be very tempting to the brethren of Brigham Young and Joseph Smith here to take on at least four wives. Can we stop 'em from taking on a dozen?"

Meshed, an important center for agricultural products, including opium, left the unholy signs of the smoker in our wake. You smelled it in the pulsating winds, its sickening sweetness congealing to your brain. Meshed, too, was only fifty air-miles south of Soviet Turkestan; but its real importance lay in the tomb of the Eighth Imam, Abol Hassan Ali Ebe-e-Muss'ar-Reza, commonly known as the Iman Reza. He had lived during the early ninth century, was revered by the Shia Moslems, with 750,000 pilgrims coming yearly from Iran, India, Pakistan and Afghanistan to visit the tomb. Nearby, in ruined Tus, which had been put to the sword by Genghis Khan, Ferdowsi had been cradled, and the epic he had written had taken thirty-three years to complete its 60,000 lines. It celebrated love, intrigue and murder, as well as opium and the roses of Persia's gaunt metaphysical gardens of the mind.

Meshed would receive the holier dervishing pilgrims in a few weeks, at which time the emotional flood of Shia Islamites would know no bounds, spilling over into murder, as it had done on many occasions, of straying Christians. Since I was a straying American, bogged down by a three-wheel scooter waiting to be reborn as a two-wheel locomoting vehicle, I was unhappily confronted by the malices of the insoluble. The night in Istanbul suddenly came back to me, the heat pricking me with a thousand thin knives. I had almost been robbed at knife point then. Why wait until August seventeenth for the dervishes to look at my blue eyes and my light brown hair? Was I "accident prone"? I wondered. I was money-order prone, still unable to cash my express checks.

But Dick Snowden settled it, loaning me two thousand rials for the plane trip back to Teheran. The consulate was also a goodwill agency, showing films to intellectual Iranians when not helping out accident-prone Americans. The Iranians were invited on Wednesday evenings to the consulate compound, with tea, lemonade, ice cream and cookies served during the showing of a documentary film offering the American way of life as an antidote to Shia Islamism. One film dealt with America and how its people became democratic citizens; and I watched the Iranian audience of one hundred melt into the ice cream and the cookie jar, wondering how much progress democracy was making with the Persian women, who tagged along, but remained silent and withdrawn. I like ice cream, too.

On Friday, Snowden took me through the ropes at the airport, where mullahs came and went dressed in their dark robes and turbans. It was the real country of Kafka, and I was his best unwritten character. For everything about my trip was now a hallucination. I was in a world of stark, sun-lit night, ready to be certified for allowing myself to get into this "exchange" mess. I had made two exits from Teheran; but I was unable to get out of Iran, being doomed to the sun's miasma and sweet Persian lunacy.

When the plane took off, the dry, brown, deadly earth of Iran tumbled up into the world, its topsy-turvy angles suddenly becoming the normal angles. I was heading back to Teheran with pilgrims in Arab dress, who looked as if they had come out of a Hollywood set for *Son of a Sheik,* with Rudolph Valentino as the handsome, cultured Arab. But I did not recall Valentino spitting in the aisles. He was not fat, either; and the well-dressed Arabs were fat and they spat all the time as they fingered diamond rings and stared at my dirty khakis. I was, without doubt, a strange-looking American. I did not wear a wrist watch, nor were fountain pens protruding from my shirt. I was a character on a scooter, suddenly air-lifted; a pilgrim bound from Somewhere to Nowhere.

I was back at Mme. Pars', with Friday being the Sunday of Teheran as far as government offices went. Everything was closed down tight, including the Vespa agency and the American embassy. I was in the middle of a muddle and an "exchange" nightmare, my sense of humor somewhat battered by this time. I was also sick again, for my head had gone to my stomach. But Mme. Pars' genial madhouse equaled my temporary lunacy, so that I was quite at home in the muddle and the nightmare, as I sought a solution to my problem.

On Saturday, I took a taxi to the American embassy, where I hunted up the commercial attaché. He gave me a letter directed to the office of the prime ministry, but addressed to the minister, asking that he try and get permission for me to make the exchange of scooters so that I could continue my journey round the world ". . . and leave," as I said, "this madhouse for a new one. A change of madhouses would clarify things all around." The commercial attaché did not think his letter would do any good.

It was a weekend of much transportation; and before it ended on Monday, I had ridden daily in forty taxis and sipped daily thirty-nine cups of tea while waiting for some important official to materialize into a chair complete with a pen and a voice, to add his signature to a vast collection of forms that I had begun to collect at 9:00 A.M. on Friday. I was soon an expert of a sort, a part-time architect and aesthete, appreciating Persian edifices and civil service offices. The stairs curled upward, spiraling like arabesque dancers heading for minarets. I had two moods—both excited. I shouted, looking like a zany, my voice ringing through the dirty marble halls, "But this is ridiculous. Do you want Iran tied up with red tape, like France? It's only a scooter and not the oil wells or Abadan!"

If red tape was evil, the red tapers were demons, at least. But I was a bull, charging through Teheran. I had a way of looking humanly sad when it was essential, or bloody mad when that became my human-animal role. One sympathetic official hit upon a way of circumventing the taping processes by asking, "Why bother changing the whole scooter; why not just change the motor?" When I tried to tell him that I wanted a scooter with two wheels instead of three, he decided that I was not a Persian after all. Did not three wheels look better and richer than two?

I finally reached the police department, the section controlling the issuing of new plates. They agreed, giving me another form. This one was addressed to the Customs, begging them to give me permission to take a two-wheel scooter out of the country.

The Customs man was also sympathetic. He was armenian, as well as a bull type, too. We plotted our campaign together. We drank seven glasses of tea and agreed that William Saroyan was a genius. He translated my American consular letter into Farsi, telling me about the virtues of the language.

I told him that Dante had an easier time outlining Purgatory; that he would have added a canto had he lived in Teheran, circa 1956.

Now it was a game and I suddenly became calm, content to play at it, to see how the bulls of the breed tossed aside the sheep and the lions in the Land of Redtapia. Now I hated lions as well as shish-kebab lambs in legal clothes, looking sheepishly at me as I stormed through their chancelleries and embassies of baa-ing confusion. Whenever I was sent away, I came back at once. After all, I had two holy documents from the American embassy and the prime minister of Iran.

I won out at the Customs with my bull colleague giving me seven more glasses of tea. The only office left to conquer was the Iran Tourist Bureau, whose manager was a lion who liked his lambs well stewed in their own juices. He was, literally, the Shah of Tourism. Without a carnet, I could only tour in the broad spaces of my own mind.

Most likely he had smoked opium the night before. He was fatter and fuller of legalisms, insisting that everything I had done during the last three days was all wrong. I must go back, starting with the prime minister; then to the Customs; then to the police; then to—hell, as I stormed my protests.

I went. I had more letters written, all of them addressed to His Majesty, the Shah of Travel. I became two bulls in a meadow, ready to slay him in his legal tracks. I had more papers than he had, I was convinced.

It was noon, Monday. I had started my campaign on Saturday. Sunday was a working day in Teheran and I worked on. When the manager began his triple talk, I exploded into savage German, aided by the export manager of the Vespa agency, who had come along to translate into Farsi for me.

I had won, but it took 160 taxi rides and 159 cups of tea. I had a new carnet, new ownership papers, my old plate, and a new two-wheel scooter. I emerged into the bright sun of Teheran ready to celebrate myself; the Crown Prince of Persian Redtapia.

We celebrated. The translator, an Iranian army major and I went to the opium district of Darvazeh Ghazvin. We scrambled through narrow streets, rounding alleys and back streets like characters in search of an opium den. We found it, at last. The building looked discreet and respectable from the outside. Inside it changed. We reached the first floor, which resembled a Caligari setting in its angular lines. A voice directed us to a back room, and the man behind the voice soon directed us to the pipes.

He sat on a stone floor preparing the substance, heating it over a brazier. The crying tones of a small child invaded the room, coming from the balcony, with the mother not trying to quiet the child. Apparently the crying child covered up the smoking and the men's voices as they began to see blue worlds in their new sensations. I wondered how the mother managed this horrible arrangement emotionally, trying to cover up her husband's nocturnal occupation.

It was a world of semi-savage consent, creating a stupor-like corruption that grew with the hour, the sweet smell of it flowing out with natural ease. It was watered by small voices talking into their dreams, in the crying child's backgrounding, in the way the man at the brazier sat as he prepared the two pipes. I, being the interested, smug watcher rather than the doer, refused to smoke.

The translator was saying that he had been a communist while working as a bookkeeper at Abadan. He put the finished pipe aside and became intimate about his past. The major was cynical and indifferent. He wanted his pipe three times a week, convinced that he could turn away from opium any time he wanted, even after fifteen years of it.

"The communists exploited me and my youth," said the translator. "They had five years of my youth!" he shouted.

The man arranging the pipes asked him to quiet down. The baby raised its voice and the mother told everybody to quiet down. Did they want the police?

It was illegal, said the major, though seventy-five per cent of the population smoked it for dreams. "I smoke it because it's better than women. You can leave the pipe, but the women are always there."

"They exploited my youth," moaned the translator. "I finally wrote a letter to the newspapers declaring that I had broken with the communists, and I was one of their strike leaders at the refinery." He was smoking again. Ten seconds later he picked up the thread: "I expected them to kill me, which they've done very often to key members who left them and denounced them. Now I smoke to find peace. Please give me another pipe," he begged. "I lost my youth to them but I can't find it in this stupid pipe. You are better off scooting around the world and not smoking; not sleeping with the whores—just being a poet. You are always young, Harry."

"If we get caught," said the Major," we go to prison for six months. But this man," and he pointed to the man filling the pipes, "he gets three years." Then he pointed to me. "But nothing happens to you, sir. You go like you came, free on your scooter. You are only watching, not smoking."

There was a legal distinction, apparently. I was the odd man looking on, outside of it all, making some moral objections to opium as an escape. I was only celebrating a victory over red tape, but they were celebrating themselves as Persians, with a victory over nothing.

A week later I was in Zahedan, having made the run through Qum, Isfahan, Yazd and Kerman, riding in a coma of heat over the Kerman-skirting desert. I turned west at Bam and crossed the lower terrain that ran near the brown furnace, suddenly giving way to green, salty marshes. The heat became worse, with the humidity increasing the nearer I came to sea level.

It was a ride through a nonconducted tour of hell, over 1,008 miles of a scorching pit. I thought then that I had been through a grand vacation during the dysentery days in Meshed. But I was a healthy man now, with my body acting like a mechanical robot, making each day's run like a man possessed, acquitting myself before nature finished me. I had a sense of desperation, for I wanted to leave this wasteland, to quit Persia before it became a permanent part of my mental fancies or before I turned into a Persian by sheer association and psychological persistence. If life was peculiar here, I was becoming so myself, unable to throw off the increasing influences dooming me in the process. But there was a mechanical relationship on the scooter, now working beautifully on two wheels, making the ruts and potholes in the sandy corrugations exhilirating rather than panicking. Only the desert, the sun and Persian fatalism was fatal. But I must leave or I was on the low road to tragedy, I knew.

From Zahedan there was only one way to Pakistan. The road to Mirjaveh was cut by the floods; nor was the railroad working; and it would take at least a month for the floods that had inundated Iran and Pakistan to subside. The floods had come suddenly, locking in the lands in watery death, with entire towns and villages in Iran and Pakistan swimming with dead livestock. The only transport was by air. I could fly over the terrible disaster that nature, in its original plan, had bequeathed to these countries.

I went to the airport at Zahedan, convinced by the local officials that I would be in mud and water before I scootered another twenty miles. A plane had just landed with doctors and medical supplies, along with three cases of Coca Cola. The Shah was due to arrive to inspect the damage, and the Shah liked Coca Cola.

The plane's captain, Bill Nicholas, was an American pilot and teacher working for the Iran Air Lines. Two years ago his plane was the last workable plane in the line; but now the line had a dozen planes, all of them captained and serviced by Americans under contract with Trans World Airlines. They were teachers, fathers, psychiatrists and nursemaids to the Iranian personnel learning to fly over the razorbacks and deserts of Persia.

They had seen their fill of dead camels, goats, sheep and bulls on the Persian landscape as they flew over the flooding waters. To the west lay the Margo desert, running into flooded Afghanistan; to the north was the I Lut desert, then the Baluchistan plain embracing Iran and Pakistan. And with Captain Nicholas in his Virginia drawl telling me to get out while the getting was good, I was soon convinced that the way out was by plane and not by scooter.

"Stay for the night at our compound and we'll work it out," he said, inviting me into the airline bus. "Are you broke or can't you afford the seventy-seven bucks to ride to Karachi with us?"

"I can't afford it."

At the compound, the Syrian woman who cooked for them told my fortune. She fed us in the garden, then took to her cards and my fate, as if I were up on an auction block. I had a split personality, she said drily; my name itself suggested that, having a kinship to Roskolnikov of *Crime and Punishment*. Also I had very little money but I was not to despair, for there would be five hundred dollars waiting for me at the American Express in Karachi. Further, I was to fall in love within a week or within a month, the time factor being somewhat blurred in the cards. There would be an important letter for me in Karachi, too. Then she announced grimly, "You must leave Iran in the morning. It is a very bad place for you—*you must leave.*" She also told me that I had recently been divorced from an "evil woman." But she was beautiful, assured the fortune teller.

If Captain Nicholas had put her up to it, he didn't show his contribution to my fortune or my possible misfortune, should I decide to take to the road and not ride on his plane.

The cook had two mynah birds that pecked about under my feet, waiting to be inspired to speech or song. One of them apparently had listened in as the fortune cards were being read, and I suddenly heard, "You must leave—you must leave—get going, Harry," giving an excellent imitation of the Syrian woman's and Captain Nicholas' warnings to me. The mynah bird cinched my debatable fortune.

"You don't have a visa for Afghanistan," said Captain

Nicholas, going over my passport. "However, if I make you an honorary member of the crew, you'll slip by the Afghan authorities."

"But I don't intend to stay in Kandahar or Kabul. I'm going on to Karachi."

"It makes no difference, 'cause they'll still want a visa. Can you write a check?"

"I can write it, all right —" but to myself I asked if my National City Bank branch would honor it a week from then. I wrote the check and I wrote a letter asking the bank to delay payment until I got to Karachi, where the mysterious five hundred dollars was waiting.

My scooter was put on board the plane. Behind the scooter came the Hajis, the pilgrims bound for Jedda, from where they would walk or crawl the last twenty miles to Mecca. Whole villages had saved money to send one man to make the holy journey, to see where Abraham had almost sacrificed his son Isaac for his faith. Abraham was holy to all men, with the Arabs, Jews and Christians claiming him for their faiths.

I was a pilgrim, too, leaving the desolation of Iran, a country that I had clung to as if I were another prophet of Islam. But the nightmare had finally lifted and I was en route to another country, with the dire prophecies of the German artist still ringing the accents of doom in my ears.

At the Kandahar airport we heard more grief. Big sections of Afghanistan were also under water.

"The DC4 is due in an hour," said Captain Nicholas, "and you'll take it to Karachi, 'cause you're still no Noah, man. Don't take any wooden rials, and good flying, chump."

Seen from the DC4 as we flew toward Karachi, the devastation was terrifying. The Indus river had gone over in Pakistan, with each country's many rivers and tributaries rising to simulate Noah's flood, sweeping through the parched lands and drowning everything there.

A Pakistani businessman was my companion. He introduced himself with a grand flourish, offered me some candy as he tied his safety belt, then sat back to study the passengers. When the plane was finally airborne, he began to fiddle with some invoices he had taken from his inside pocket.

"I'm importing rugs from Kerman, which I sell to the Americans in Pakistan; you know, the wives of the officers."

There was more rattling. Soon he was adding to it, drawing from his portfolio several small boxes of candies, which he opened with the same grand flourish. His fingers were delicate and long; he was well dressed, Western style; and the exaggerations of his physical movements were the residue of British training, the Moslem religion, and his conception of the American approach to things of the material world.

"If they hadn't murdered Ali, Pakistan would be a great nation today," he said, out of the blue.

"Too bad," I said, suddenly remembering the politician he had mentioned. "Partition is never good. It only divides and seldom adds. You would be greater if you had remained with India as a nation within a nation; a sort of little United Nations of Asia, similar to the United States of Europe in scope."

"We are nationalists first," he said smiling and rattling the boxes of candies. "Have some, please."

I took a piece, then looked at the wall of clouds we had just entered.

"I'm nervous," he said, putting away his papers and his portfolio. "I have many things on my mind. Perhaps you can help me? Do you need money?"

"Who doesn't?" I replied, hardly believing that the mystical five hundred dollars would materialize in Karachi. I am only too human. I told him about the five hundred dollars and the fortune-teller in Zahedan.

"You must be very romantic to believe her," he said softly. "You can make that much right now or just after we land in Karachi, if you'll do something in a business way for me. Will you? It's a very small favor. Will you?"

"I'm not unromantic," I answered, going back for a moment. "I have a literary agent in New York who sells half of what I write. Perhaps he's romantic?" I pulled out my cigarettes and offered him one. "Whom do I have to kill for the five hundred? Your new prime minister, perhaps?"

I laughed, hoping this would not offend him overly. Besides, it made the boredom of flying easier if one committed a little Hollywood plotting on an international level. In my dirty khakis I must have fitted the role of a unique spy, with an offbeat accent—and he laughed all the more when I told him about the scooter. Soon he was back with his candies, telling me about his children, his business, and the wonderful Americans he knew in Pakistan.

"But what do I really have to do to get this five hundred dollars?" I asked again.

"Just take a little package with you," he said in a whisper. "You understand now, don't you?"

I almost understood, I thought. Was it diamonds being smuggled in or was it opium? But opium was available; besides, Pakistan had something just as good or just as bad, Indian hemp, which made you cool, then delirious, then deadly numb, if taken in three doses. It must be diamonds.

"Expensive glass?" I asked quietly. "And if I get caught, it will make very silly reading, won't it?" I began to visualize headlines in Karachi and in New York: "American poet on scooter nabbed as a diamond smuggler," or something in *Variety* going like this, as if it were a vaudeville act I was in: "Bard Kleptomaniacs Karats in Karachi."

"I'll make it six hundred and fifty dollars. It will only take ten minutes of your time," he said again, his whisper rising dramatically.

"I'm a self-appointed diplomat, not very good at smuggling." I grinned delicately, wondering whether I was a fool or just a coward afraid to take a chance. I decided that I was neither, and I shook my head, saying, "Sorry, I don't like illegal affairs." I got up to go to the toilet, leaving him a way out. When I returned he was sitting up front, talking to another Western-style passenger—and I wished him luck.

Chapter 6

THE DIVIDED LAND

KARACHI WAS SAD, filled with Moslem refugees who had fled partition and Hindu India. A million and a half had flooded into Karachi, camping out besides the rich homes, hoping to embarrass the owners into a state of charitable grace. They lived in grass huts, in cellars and gutters, making temporary homes from discarded gallon-tins and packing cases. But it was no longer temporary for them, for they were bringing up families in their new world. It was as permanent as nature, equaling the human flood brought about by man and much more devastating than the ravages brought about by the Indus and the Tigris.

I wrote the following in my daily journal: "The rain is falling again. Yesterday one inch fell in three hours; and from the look and the sound of it, it will be three inches in three hours today. My clothes are damp and do not dry; my bed is the same, for it's a wall of water everywhere, sweeping in with the winds through my latticed windows at the North Western Hotel, where I am staying in Karachi. A few more hours and all of Karachi will swim into the sea. At Kotri the Indus has risen almost twenty-three feet; and so has the Chenab. The Lahore-Multan-Quetta road is under two feet of water, and this is the road I am to take to get to Bombay, via New Delhi.

"In the city the refugees live in seas of water; and only the tall, indifferent camels, hauling the wagons loaded with firewood, appear to enjoy this. They move by, heads poised like Gods or like dignified plenipotentiaries, trying to bite you as you saunter by.

"The people of Karachi have the most eccentric forms of transport, from pedicabs, taxis, three-wheel scooters fitted up as taxis, to droshkies pulled by camels . . . and when the rains come tumbling down, the Pakistanis take off their shoes and take to the puddles. The Westerners stand under trees yelling for taxis and cursing the rains and natural calamities."

I had lost my sense of time, if not several of my senses, with a month having gone by since I last saw a newspaper. I was shocked to learn how many calamities were being prepared for the East and West, with Egypt and the Suez leading for political honors. Every Englishman I met thereafter wondered how soon he would be going home to change from mulfti to khaki, for Pakistan was pro-Egyptian, if merely because of its normal Moslem sympathies. The newspaper, *Dawn,* which became my informant, agitated angrily against England and colonialism; and it looked like war, at least in the newspaper; and it looked like war to the English civil servant, Mr. Charles Wells, who was staying at my hotel. Mr. Wells was a short, fattish, worried man, who was most un-English in his rhetoric and social attitudes, the last of the English civilians in Pakistan and India. Mr. Wells had a unique job, especially so in a time of contracting colonies. Oddly, he was concerned with expansionism, or how to enlarge United Kingdom offices to fit in more chairs and tables for civil servants complaining they had no room to lean on their elbows.

"We can't afford a war at all. We're broke and we can't get any broker," he said sadly over tea and what passed for crumpets.

"Who can?" I managed, getting privately philosophical. "The profits of war are strangely not profitable any more. I'm broke, too, and I'm not a country." We laughed and ordered beer.

"I've been in India for more than twenty years," said Mr.

Wells. "If I had some decent concern for myself, I would have left in 1947, during the exodus. Now I feel useless and very much unwanted. We're the uncles who overstayed the dinner invitation by one hundred years; and now it's only the crumbs of commerce or the crumbs of colonialism left in the antiquated breadbasket. How long are you staying in Karachi?"

"A few days" I answered, drinking the hot beer. "How's life in New Delhi? Is there anything besides cocktail parties for the diplomatic corps?"

"I went to one before I left," and Mr. Wells, in a moment of casual discernment, told me about Han Su-Yin, the literary doctor who wrote *A Many Splendored Thing*. "She is not very pretty at all, perhaps even cold. I read her book, but found little in it for me. Can't say that the Eurasian complex is very appealing, or the ladies indulging the complex are."

I had heard another indulge the Eurasian complex; for I had recently been divorced from a wallowing, beautiful member of its fateful family, after swallowing its intimate poisons in increasingly larger doses. It was a brutalizing world for them, though the pain was self-inflicting, stemming from their self-conceived torture chambers. They lived in two shadowy worlds but were part of neither, at least in their curious analysis. They were show pieces on exhibition. They dressed as Eurasians for entrances and exits, but their minds were as Western and as materialistic as pawnshop owners on Eighth Avenue in New York. As Mr. Charles Wells talked on, I thought on, remembering seven years of dualism. We had equaled each other; then erased each other and loved each other in all the conceits we could develop for each other. But I could no longer cut my throat and neither could she. The nightly sweat and nightmare of her strange willowy mind, living in a psychological, personal prison, soon derailed and destroyed us. If there was love, I gave the relationship love. We were admired, with all our friends insisting that we have children, "who would be so beautiful." It was a cliché we suspected when we were first married. It became an ugly bastardy of fractionalization and competition, though I never competed with her.

At the American Express office I found two checks for me, together adding up to five hundred dollars; and I felt as if a mystical ritual had been performed for my private needs as I recalled the Syrian fortune-teller in Zahedan. The money was real enough; the force that had sent it was equally real. My agent had sold two stories and the magazines had paid for them upon acceptance; but it still had a supernatural air about it.

The cheapest, shortest route to Bombay was via the S. S. Sarasati, a three-thousand-ton coaster that made the five-hundred-mile run in thirty hours for only twenty-two dollars. By scooter, there was only the long way around, via the Northwest Frontier and New Delhi, or a distance of two thousand miles. I had asked the local automobile club if I could go directly south to Bombay; for on the road map it seemed only a two days' run.

"There is no road to Bombay, sir," said a young clerk. "You must go by boat or through New Delhi." Either the road had been cut, or the Moslem-Hindu separation of Pakistan from India did not include a short route between both cities. The borders between the countries had been stretched to the ridiculous limits of geographical nationalism, which my scooter would have no part of.

Going second class meant that I would not sleep on deck, which the passengers traveling third class did. Nor did they get food, which they packed in wicker baskets. They packed everything, in fact, staggering onto the ship with enormous bedrolls, which they promptly unrolled over the decks. I packed my scooter on, tied it to a stanchion, then found my stateroom, as well as a fifty-year-old Irishman, an old India-hand, who was sharing my cabin. Mr. Quinn was very thin. Mr. Quinn was going home after twenty-six years of railroading in the Punjab. Mr. Quinn had the cabin filled with eleven wooden cases of artifacts he had collected during his years away from Dublin.

"I wondered who you might be," he said. "When I saw your last name, I thought you might be a Russian spy or a diplomat; though what would you be doing on this tub? But when I saw your first name, then I thought you might be an American."

I had unpacked a bottle of cognac, which usually settled my stomach at sea. I offered Mr. Quinn a drink. He refused, his eyes looking tragically at my glass, as if he were going through a private drama of some sort.

"I can't drink any more," said Mr. Quinn. "It seems that I had an enforced twenty-six years of it. Now I can either live a dull life without it or I can die from it. I drank up the money I didn't send home or I couldn't send—and now I'm sending myself home."

He looked eaten and beaten, as if time and alcohol had done a thorough colonial job on him. He had run away from Dublin and taken to the railroads of India, a typically romantic type of Irishman, who needed larger worlds and darker continents, or more poverty than the Irish could claim for Ireland, as well as a sense of regal indiscretion, allowing him to do as he wished through the sheer distance from his homeland.

His thin, alcoholic face was worn and aged; and he was very much the silent man, suffering all sorts of mental and psychological hangovers. His conversation was abrupt and awkward even when he talked expertly about India's political state of affairs, but only when he mentioned his two daughters did he seem to lose his halting, awkward speech, and a sense of poetry would come over him. "I want to be with my family now, even if it's only a grass-covered hut in Limerick. I feel like seventy and I look it, Harry."

The sea was very rough and the side canvas had been put up to protect the third-class passengers sprawling on the decks. The monsoon was in full swing and the rains lashed through the openings in the canvas, spilling on the women and children huddling together, with the children crying each time a gust of rain splashed their colorful costumes, now dripping with sea water and rain.

"The monsoon will end this month," said the Indian purser

who had helped me lash the scooter to the stanchion. "But you must go by the way of Poona, if you're going to Ceylon. The road is good and the scenery is very beautiful. Besides, Poona was once "forever England," so you should see the capital of her past military colonialism."

"How about going via Goa?" I asked.

"That depends on the Indians or the Portuguese, for the frontier between India and Goa is closed. Why, we can't even trade with the Goanese. If we did, we wouldn't be allowed back into India. Goa's an unholy place now," laughed the purser, who seemed to delight in making his ironical statements sound even more so.

I remembered Goa, for Saint Francis Xavier was entombed there in the same position as Lenin and Stalin. It was holy with politics, too. The Portuguese and the Indians were warring over its tiny spaces, with the Goanese anxious to retain the Portuguese because Goa had a higher living standard than any other part of India. With Goa incorporated into India proper, the high standard would disappear. Nationalism would give them nothing but a warm emotion to warm their homes.

As we passed through Karachi's bleak harbor, Mr. Quinn, who was now my local historian, pointed to a little series of islands jutting up. "That's Oyster Rocks, where the Munsakhani murder case started. It seems that before World War II, a Mr. Munsakhani took his pregnant wife to the Rocks for a picnic and did her in. Lots of people have been murdered there. Why, even a priest got himself killed by falling down the rocks, and all he wanted to do was to prove the innocence of Mr. Munsakhani."

It sounded like another version of the murders on the Galápagos Islands, which I had once visited on a schooner. Then it was an over-romantic countess, who allegedly killed two of her gentlemen lovers, without even a priest to intervene. I wondered about things spiritual in the world of mayhem, when Indian gentlemen and German countesses took up the gun to prove how romantic they really were.

The ship pitched heavily throughout the night, its three

thousand tons half out of the water or heeling to port, ready to keel over. If there was cause to be alarmed, the sailors and officers did not show it, though the third-class passengers betrayed their excitement with every pitch of the vessel, their voices lamenting whenever the swooping sea appeared to engulf the vessel.

Unable to fall asleep, I managed to dope myself with pills that erased the earth as well as the sea, with a brief visit to Nirvana-in-the making; that is, until a chair went hurtling toward my bunk and landed on the bunk below me, hitting Mr. Quinn on his noggin. Now, much awakened by the sea and accidents, I found my cognac bottle, took two gulps, offered some to Mr. Quinn for medicinal values, and read Indian lyric poetry until morning, wondering whether I was undergoing more Iranian hallucinations.

"That's Anjar," said the purser over breakfast, pointing toward the nearby shore. "Last month there was a bad earthquake there and more than one hundred people were killed. Houses collapsed even as far as Bombay." He paused, changed his conversational tack, then added, "We used to call in at the Cutch, but we can't call in any more, because the sea's too rough. For twenty new passengers we might lose the ship."

He was a herald of calamity on a local scale, seldom digressing from his choice pickings of calamitous events. "Why," he continued, "last month we had a peculiar thing happen. A woman on board gave birth to a baby she did not want. She was a very modern woman, apparently. She was alone in her stateroom, without a doctor, without anybody. You see, we don't carry a doctor. The captain said the woman was unmarried; that she was merely using the ship as a self-appointed delivery room; that she got rid of the child because she did not want the child."

"Did anybody see her throw the child overboard?" I asked, wondering if this were another sort of romantic-tragic Indian tale intended to amuse me after a bad night.

"Of course not, or we would have had a murder charge leveled against her. She came on looking very fat, obviously much pregnant, and she got off looking nine pounds less in weight, which is what the child must have weighed. And even if she did not eat our very good food, she could not have lost so much weight overnight, could she?"

It was quite a mystery, item for item and weight for weight; and happily I decided not to try and solve the problem that morning or that year. I did, however, plan to get more information, if only to do a story about it for one of the magazines that wanted off-beat stories. The magazine's editor wrote me: "There must be large red-light districts in Istanbul, Teheran, Karachi, Bombay, etc. Please get me the local color, as my readers who are mostly young G.I.'s, want that sort of purple coverage. I can pay you two-hundred dollars for 2,500 words and something extra for pics . . . and I can use all that you can uncover."

Chapter 7

TOUCHABLES AND UNTOUCHABLES

MY FIRST VIEW of India was of its intense poverty, then its political drifting. It appeared to be waiting for some dictatorial magic, for the strong man to buttress up the falling social structure. Nehru had taken on the role of the benevolent father for ancient Mother India, but there was riot and murder in the streets, especially over caste and language problems. India was straining to get some order in its Babel of tongues. How many national languages could the schools teach? English was still the second language, but there were other claimants for the second language and for a third and a fourth one. The claimants set up political demonstrations, and a summer of slaughter was spreading from Bombay to Ahmedabad, its political savagery taking an enormous toll. Nature was not neutral, either, for the Indus floods were still ravaging the land, and it was a depressing spectacle that greeted me in Bombay. All the newspapers ran scare headlines, investing every killing with greater and greater alarm.

But India had progressed industrially and socially. I had been to India in 1947, when the caste system was still in its heyday. But now, nine years later, Indians were still killing each other; and though the caste system was slowly "withering away," as the Russians used to say about Western capitalism, it was still blighting the land of Buddha and Gandhi. The real withering had been among the English colonials, who had seen their former economic paradise taken over by Indian entrepreneurs. Prohibition, too, was causing another sort of disappearance, that of the drinkers, with the remaining Englishmen paying a dollar a slug for whiskey and beer costing eighty-five cents a bottle. Those whom I met at the semi-private Ambassador Bar complained bitterly about Nehru's plot to dry up their drinking habits.

Though even an Englishman can't live by bread or drink alone, the Englishmen I met were doing the latter, with the memory of colonial India prodding them on brutally; and the more they drank the more their ambivalence showed up in social and political matters.

The Australian artist Roy D., whom I had known during the war years in Australia and later in Paris, was now no longer the free-wheeling Bohemian concerned solely with his art. He was the art director for a gigantic international firm, having safely committed himself to better living and more drinking, his past social conscience giving way to the virtues of economic certitude in his newly discovered middle-class life. He talked more about art, but painted less. He got fatter and sounded less desperate about money. In the company of the Englishmen he seemed to polarize his new values by saying, in a moment of joy, as if this were also an apology for his new physical grossness, "Harry, in five years I'll have enough to retire; then I'm going back to Paris and back to the easel."

The talk about the Suez and Colonel Nasser seemed less ominous; it gave Roy D. a moment of aesthetic grace amid the three ex-colonial Englishmen. One of them was a doctor who specialized in removing a few female internal organs, making a fine living via the new cure-all, which someone had dubbed "hystericalectomy"; another, a director of an oil company, found it juicier siphoning Scotch on the rocks, much as if he were bringing in a gusher; his companion, a pretty English girl with nothing to do, wondered if she had better not go back to England before her chances of a good marriage disappeared completely.

"Perhaps, if we stay long enough, we'll discourage Nehru from bringing in socialism, or communism, or what-have-you," said the director.

"They have an intelligent civil service, thanks to us," said the doctor. "Now they are very efficient at making riots."

"I saw an untouchable killed the other day," said the girl. "It happened so quickly, the police couldn't stop. I saw the man kicked to death, and how horrible it was!"

The doctor bought her a drink, then said, "You need something to quiet your nerves, darling."

"Like a husband, no doubt," said the director. "Would you consider me, my dear?"

"Are you proposing or propositioning?" asked the girl with an acid smile. "I'm quite untouchable on propositions. Proposing?"

"I might, my dear. She'd look very good in a drawing room, wouldn't she, Harry?"

"Drawn and quartered in a latter-day ex-colonial drawing room," I answered unthinkingly.

My humor seemed wrong, and the Australian bought another round of drinks to ease the tension. We talked about decadent Paris for a few minutes, then off I went to the Regent Hotel on Ballard Road, somewhat frozen out by the director.

Bombay's streets made me shudder for Indian humanity. The blind, the crippled, and masses of normal-looking beggars were at every juncture. They had voices, elbows and special antennae, proclaiming that I was "a rich American capable of giving a great deal of baksheesh." They vied with the snake charmers, the fakirs and the acrobats; for each of them was a performer playing for odd coins from sympathetic strangers. A couple and their two-year-old boy were performing a specialty act. As the child was thrown into the air, the father did leaps and tumbles while awaiting the child's return to earth, straightening up in time to catch the falling child. Once the child came down as the father was finishing a whirl, and the child fortunately received only an abrasion instead of a fractured skull.

Across the street from my hotel a young woman sat scratching away at the lice on her body. She had been there for three days and nights, through the heat and the monsoon rains. Her bed was a burlap bag; her belongings, a large kerchief filled with odds and ends. She had only one other gesture—the right hand going out to every passer-by, and her mouth uttering only one word—"baksheesh." Every Indian language was lost on her, with "baksheesh" her only word for hate or love. Her right hand, the rain, the burlap bag, and "baksheesh" were still modern India.

Bombay was a mass of colors playing violent games of contrasts, with large dark women, their foreheads decorated with red and black designs, flashing by along Marine Drive, their saris embroidering the humid air and the adjacent sea. I had the usual assignment from a magazine in New York, the rope-climbing trick; but no one was climbing ropes that afternoon.

I met Shaun Mandy, the editor of *The Illustrated Weekly of India,* and we talked about the sort of problems an Irishman had in editing an Indian pictorial magazine. What could he, a foreigner, criticize in India's political and social pageant? Everything he printed was an exhibition piece to show the country growing, fusing the old culture with the new in the making. His editorials were subject to the pressure that Nehru's tremendous prestige exerted, which consciously or unconsciously silenced most would-be-critics.

Shaun Mandy took me to lunch, bought four of my poems, handed me sixty rupees, and introduced me to young Dom Moraes, the son of Frank Moraes, then the editor of *The Times of India.* Dom, at eighteen, showed the serious poet's flavors and toxic ingredients, as well as the flair that separated the good poet from the hordes of versifiers who had nothing to say and managed to say it very well. But Dom, good poet that he was, was a bad student, having barely managed to pass his entrance exams for Oxford. Accordingly, he said, "They say that I am ignorant and know nothing of Romance languages, including Latin; that I'm hopeless in the sciences; and that the only thing left for me is English, and so I'm majoring in that."

According to Mandy, who saw all things through an editorial eye, the world had some obvious evils that were not always political. Opium was a teen-age evil, along with other narcotics; for he had read American statistics about the jungle of sensations haunting the schoolboys of America. In India it was a variation on the same theme, with teen-agers using Bhang, the green leaf found in the Sind of Pakistan. Bhang was ground with sweets, then mixed with a sweet drink. One glass made the drinker cool, but sexually potent for six hours; a second made the drinker tell all the truths he knew; a third put the drinker into a frozen coma for forty-eight hours, unable to feel any sort of sensation, including sexual desires.

Bhang could have a bizarre future in America, especially if blessed by Madison Avenue's ability to see anything at any time. Even dull political campaigns, through the subtle use of Bhang, could wind up in great orgies of rhetoric; in any event, the listener could always take another drink, go numb to the polls, neutralized but voting for Bhang, the candidate that offered everything and delivered it, packaged for a comatose conscience. And since advertising sold America to the Americans, it would eventually do the same to India. But advertising was an infant industry in India, awaiting American methods in sales techniques.

When I met Nissim Ezekiel, the editor of *Quest,* the new literary organ published by the Indian Committee for Cultural Freedom, I saw the infancy of sales promotion desperately trying to hold back India's furtive call to communism. Ezekiel was an intellectual using satire to expose communism; whereas communism used the obvious tools of poverty to sell the miseries of communism to the poverty-ground, million-headed Indians, seeking a way out of poverty's face and body of death.

Ezekiel was an Indian Jew, stemming from the Jews who came to India after the Dispersion. He was light skinned, racially conscious, intellectually hypersensitive, critical of Nehru, and more so of Krishna Menon, whom he accused of acting as the the midwife for Indian communism. "But he's extremely shrewd, a man who has acid rather than blood in his veins," said Ezekiel. "Some say that he lies, yet that his lies always sound like the truth. He is that facile."

"Perhaps two glasses of Bhang would get the real truth out of Menon," I suggested. "Does Nehru know what Menon really thinks?"

"It hardly matters, for the aftermath will be the same; appease Chou En-lai or present Chinese communism as a respectable, decent, independent movement. What contemptible frauds!"

We were walking down Anne Besant Road, having at one time during our walk crossed Mahatma Gandhi Road, thus feeling two sorts of metaphysics dragging after us, with their ghostly literature echoing in our wake; for Bombay, like Paris, revered spiritual-minded apostles, naming streets after them in the hope that the pedestrian-minded would find a state of grace. But we found more beggars, with Ezekiel insisting that too much metaphysics was what modern India was still suffering from—the metaphysics of Nehru.

Nehru, despite his awesome dignity and past sacrifices, was a small man intellectually, having more than enough vanity to puff himself into a peacock should he care to change his human form, or so said his many critics. Those who really opposed communism found him ambivalent, a man who could not be said "no" to, though it was agreed that he would make a poor dictator. Once in a spirit of self-criticism Nehru saw himself as a dictator, and his confessional regarding his dreamed-of role did not disturb his avid followers, who never found him wrong even when he found himself wrong.

When Nehru addressed mass meetings, he lectured down to the masses like a benevolent parent, insulting his ignorant audiences or wheedling them into line—Menon's line. Not oddly, most of Nehru's cabinet distrusted Menon, for it took three years of cajoling on the part of Nehru to convince them that Menon should be a minister without a portfolio. Some of them threatened to resign; but since none had, Menon's voice was the voice of India.

"More women than men commit suicide in India," said Jean, a French photographer who lived at my hotel. "Everybody is unhappy here, so they kill each other or they kill themselves." But when crowds followed the much-bearded Jean, who looked bizarre in his beret, he threatened death to all and sundry as they peered into his camera and kept him from working.

"Are you very unhappy?" Jean would ask some well-dressed Indian, who ignored Jean's protestation that he was interfering with his photography. "Do you not have a wife and a home? Do you not have at least one mistress? Or does your mistress find you very boring? I am not your mistress and I also find you very boring, Monsieur. Perhaps you should like to try dying for a little bit. It will not hurt very much, Monsieur," and Jean pulled out his shoulder-holster pistol, pretended to cock it, and in his unsoldiery way he aimed it at the temple of the man.

The children in the large audience found this more exciting than the American westerns; so wherever Jean went he had a natural audience, much as if he were a French Roy Rogers. But being large, fat and bald, Jean was out of character. When he was not out photographing, he was seducing, after a fashion, any number of Anglo-Indian women; and when his Gallic emotions really stirred him to thoughts of emptying his pistol, he went out into the country and shot at the trees.

My hotel on Ballard Road had suddenly been transformed into a Catholic church, for the real church had caved in during the monsoons; and every Sunday saw the dining room as well as the foyer transformed for Catholic services. On Sunday the Goanese waiters dashed about with trays, rushing in and out of the bedrooms, hoping to induce some of the guests to attend the services, so that breakfast, if not the services, became an anarchistic affair, physically and spiritually.

My immediate neighbor was James S. Tong, S.J., who called in when he heard that I was a fellow-American. Father Tong was an American from Kentucky and not a Chinese, despite his name. He had been away for more than sixteen years, moving Jesuitically around India as a pinch-hitting preacher. He complained about India's new attitude toward missionaries, though Nehru had recently allowed two more Jesuits to enter India providing they promised to teach nothing but agriculture, thereby planting no grapes of religious wrath. There was only one religion-in-the-making in India's fertile earth—and the wrath it would sow would have nothing spiritual about it, when its red blossoms finally came up.

But India was also bursting with all sorts of dramas, especially in its films. I saw *Jhanak Jhanak Payal Baaje,* a lavish dance-drama done in overlush colors, overripe in its setting and often terribly embarrassing in its direction and acting. But the two dancers, Sadnhya and Gopi Krishna, were excellent when they performed Indian classical themes. During the intermission, or ninety minutes after I had entered the Opera House, I left for the religious solitude of my hotel, or just in time to hear Father Tong say, "What India needs is better religion," to which I could only say, recklessly, "Or perhaps none at all? Look what religion has done to India." Father Tong shook his head, uttered a prayer for my sudden heathen stance, and we bid each other a spiritual goodnight.

The next morning Dom Moraes, in his chauffer-driven car, took me to see the second century Kanher Caves on the outskirts of Bombay, which the earlier Buddhists had used as a fortress in defense of their religion. But the monsoon hit us as we approached the hills, with Dom apologizing for the weather, for the mucky red roads, for the absence of signs and almost all of physical India. On the way back, we passed Fatempur Sikri, and Dom pointed to a well off in the distance, saying, "The young boys of the neighborhood will dive into that shallow well if you throw a coin. It's a trick, of course, or they would bash their skulls. I wish that I could do something like that."

We passed a holy man wallowing about in his saffron sheet. Seeing him, Dom exclaimed in boyish glee, "Look, now they wear wrist-watches. It's a bad sign, for the spirit is being refrigerated by concepts of time. Soon neon lights will be decorating their temples."

Later, Dom Moraes said over tea, "In Poona there is a poet named Borka Mamatmaya, who is writing a heroic poem, an epic of the life of Gandhi. He gets a hundred rupees for every hundred couplets that he writes, which is paid to him by the government. He writes a hundred and fifty couplets a week, so he's living fairly well though he's fast running out of material. But he's a good poet and very sincere," and Dom laughed, as if the economic relationship of so much per couplet was too ludicrous for our times. Getting paid for poetry, especially for epic poems, was just too much for his sensitive mind to contemplate, and I wondered how Auden, Frost or Sandburg would manage if they were asked to write an epic poem about Roosevelt at a dollar a line.

When Frank Moraes, Dom's father, returned from his junket to China, he became more critical about Red China's political straight-jacket. Being courageous in his editorials in *The Times of India,* he tried to correct the errors of Nehru and Krishna Menon. When Nehru went to the Soviet Union, the Russians refused Moraes a visa, considering Moraes not respectable enough for Soviet fellow-traveling respectability. I wondered how I would approach the question of Nehru's stewardship when Wednesday came, for I was to have dinner with Frank Moraes that evening. Further, Frank Moraes was soon to have his book on Nehru published by Macmillan; after that he was going to the United States to lecture on India and/or India's neutralism, its planned economy and the communist threat in India.

John Strachey, the noted British Labor Party expert on the decline of capitalism, had just spent two months in India surveying the second Five Year Plan and its progress. Before leaving he left an eleven-page summary of his findings; if the Plan failed, India would turn to communism to achieve the merits of the Plan; but his sad opera of fact and fiction did not seem to disturb too many people. Indians either loved rioting or they accepted dire *pronunciamentos* with equanimity, and that included the dispossessed Mahrattas, who had once owned Bombay state but had lost it, commercially, to the Gujaratis.

The much debated caste question was not part of Strachey's report on the forthcoming Brahmin day-of-doom; the report dealt with industry and social engneering, not with the commercial morals or the social and economic power of the industrious

Gujaratis, who, though only seventeen per cent of the population of Bombay, literally owned the commercial city. They were skilled merchants, abstaining from many modern human activities, including the eating of meat and fish and killing; for should an ant cross their dinner table as they dined, the ant would get its share of their worldly goods. But when the Gujaratis and the Mahrattas clashed, there was murder and mayhem left in the wake of the live ant.

The once noble Mahratta had not kept up with commerce and progress, having abdicated his past rulership; now he was in the dregs of Indian society, the day laborer, not the noble prince; and the riots, though apparently related to the question of bilingualism, had their real source in commercial hatred or the usual envy of the dispossessed against those who owned the stores and the industries. During the days of the Mogul kings, the Mahrattas had lived in warring splendor; but there was nothing noble in the riots they engineered; nor was India the winner.

According to Nissim Ezekiel, Indian politics, like Indian thinking, was in a state of Indian confusion—or normal; Nehru's cabinet members were merely rubber stamps acting like expert politicians; India had more yes men, collectively, than all the combined countries of the world; there would not be human progress in India until the caste system question was solved; it was futile to talk of an Indian foreign policy, for the country was drifting into a policy out of sheer lassitude; and if India was really and objectively neutral, the West would have an ally. But India was threatening poor Pakistan and grabbing Kashmir. India was endorsing Nasser in Egypt and cultivating all the social approaches to Chinese Marxism, which was hardly the socialism of the British Labor Party.

Nissim, who loved the stage, had helped to launch a theater that put on T.S. Eliot, Giradoux, Christopher Fry, as well as other playwright-poets; but Ezekiel saw little future for the literary drama and he challenged India on every level of its aesthetic consciousness. India worshipped death, not life. It was merely a marble repository, living off a nationalistic arrogance. Its great statuary was symbolical of its inner totalitarianism.

"There is no real theater today in India," said Nissim during dinner, offering me a glass of red wine. He did not drink himself and had bought the bottle to please my Western tastes, and I toasted silently to things bizarre in a world of strangers and strangeness, where the intellectual was often as stiff-backed as the things he deplored, unable to democratize his spirit in the give and take of ideas. I listened to his comments about Indian films, for which he apparently had a special loathing, equaled only by his attitude toward Krishna Menon.

"When the talkies came," continued Nissim, "the Indian live theater stopped talking and acting, literally quitting as an art form. You might just as well give up Shakespeare or call him old hat, for it amounts to the same sort of aesthetic suicide. Now all that we have left is an amateur theater, with a few semi-professionals keeping the memory of the stage alive. As for the films, especially the mythologicals—well, you've seen *Jhanak Jhanak Payal Baaje*—so you've seen them all. But we've done two good films, however: *Father Panchali* and *Khajurahe*."

After dinner Nissim talked about the Indian dance, saying it was the only thing worth seeing after the sculptures on the temples. A few months back, the American, Jean Erdman, had danced in Bombay, and Nissim was still rapturous about her as an artist and performer.

"It had the breath of modern life, Western life, if you please. One saw fusions that enriched the modern spirit, giving us a clue, perhaps of the psychological depths that have made America great."

Jean Erdman's husband, Joseph Campbell, had edited a Bollingen Foundation book on Indian sculpture, and together they were Americans on exhibition, the Americans of the arts going abroad for a season of democratic tête-à-tête. But the Americans going abroad on cultural safaris were much behind the Russians in showing off our best artistic foot, which usually ended up in our loud mouths. We appeared to lack the challenge, as if we were smugly satisfied in exhibiting our American way of life, which we agreed was good for everybody. An unqualified self-righteousness had entered our once very individualistic attitudes, and money did the talking instead of our vigorous arts.

Nissim's little daughter, Kavita, ended our conversation when she recited "Jack and Jill," falling down to illustrate the poetic jingle. *Kativa* meant "poetry" in Hindi, and I wondered how far America had fallen down the hill with its many pails of foreign aid en route to save the world in the process.

Frank Moraes phoned in the morning that he would send his car around at dinner time, that we would dine at the home of his publisher. It was a lavish, modern apartment with Western-style functional furniture, its floor-to-ceiling bookcases filled with literary and historical American books. Moraes drank Scotch like a working journalist at ease. He had a gracious, warm way, a fine voice, an accurate mind, especially in political matters. And like all good editors he had a story to go with the moment, which stretched into six drinks before dinner, during which time Moraes mentioned his projected trip to the United States. He was to spend ten days at Columbia University, then enter the American hinterland on a lecture tour, though nothing had as yet been planned; and he was leaving by plane within two weeks.

"You need an agent to book your lecture dates, to take care of all the details," I said, when he seemed confused as to the cities he was to visit.

"But agents take fifty per cent; they also make you pay your plane and train fare," he said. "What would be left for poor me?"

"You still need an agent to organize your trip. Unfortunately, this is American efficiency, which all of the world wants without the American label. English lecturers love it—and it pays so well, especially when an Empire is passing away."

"We can use some of your efficiency here," answered Moraes in his gentle way. "We are organized for bedlam mostly, not for practical pursuits."

"Do you plan to write for American papers?"

"Lester Markel of the New York *Times* cabled asking me to write for him; but he didn't say whether it was to be about the United States or about India. I suppose it will be about both countries; though what can I say about the United States after a two-month visit?"

But he had found enough to write about in Red China after a brief six weeks; and what the United States would reveal, particularly regarding our internal Negro problem, would define for him the democratic process of the American within his own borders; and like so many Indians, he would at times be mistaken for a Negro, especially if he went about in Western-style clothes. The Ceylonese poet, Tambimuttu, who used to wear Bond Street clothes when he lived in London, wore completely Indian garments in New York.

During dinner, Moraes said that the entire circulation of all the English printed newspapers and magazines in India was slightly over four million, including the religious magazines. His own paper, *The Times of India,* had a circulation of only one hundred thousand, and Hindu papers not much more. The Indians preferred movies to books or newspapers, and when television entered India, they would cease to read completely.

His publisher, who also published textbooks because they were safe investments, said that Indians read Indian junk along with American junk; that all I had to do was look at the ornate covers on Indian paperback books. "The colors are richer and riper," he added, "but the contents are utterly stupid and trivial. Anyway, we are still a nation of illiterates here, so that perhaps is the answer."

"The Americans are not," I replied, accepting his critical cue. "We are literate enough to read several hundred million comic books, along with an occasional classic. We are very inventive, we Americans."

"Literacy without a sense of literary values is not literacy," said Moraes. "But I suppose that's what mass education does; it spreads a little learning among the many."

Later, somewhat entranced by my scooter trip, Moraes

wanted to know whether I had some very private reason for this literary safari on two wheels. Was it for my physical or my mental health? Was it a cheaper way to travel? Did I see more and better from a scooter's view of the world? Or was it an adventure for the abstract man inside? Whatever it was, he envied me my months of freedom and my lack of a responsibility to a deadline.

But I had a deadline for the man inside. David Thoreau had Walden and the pastoral setting before industrialism swamped his heartland; he wanted to *know himself*. Whitman had the arbitrary prose-poetry era, with the straw-stuffed inner man delivering orations about the forthcoming democratic age, making himself the cadenced rhetorician of the prattling ego. Melville had found a mystical entity and a fury that went with his myth. And Roskolenko, soon to be forty-nine, was nobody and trying for nothing, yet somewhat cynical about some American facts and their universal fictions. The world was rapidly running into limbo, via neutralism, and I was merely observing some of the strange imagery en route to a Neutral Cosmos.

I left for Poona late in the afternoon, with the Frenchman, Jean, acting as my guide for twenty miles, his heavier Augusta scooter racing through the monsoon rains like a wraith on wheels. Before parting, Jean invited me to visit him in Lyons, where he ran a restaurant as a sideline to his photography. A year ago, while on a photographic expedition in Egypt, he had crossed the desert and found three English girls lying unconscious inside their broken-down car. Two of the girls died, but he managed to save the third girl, an attractive, wealthy "lost" creature who wrote Jean weekly that she loved him and wanted to marry him because he had saved her life. I remembered seeing the story in the newspapers; but it seemed odd meeting the "hero" himself; for Jean in his flamboyant way was the modern adventurer engaging with all sorts of lost girls and married women, using seduction as a science in his restaurant in Lyons, to which he was returning as soon as his agency sent him his fare back.

A few miles out of Bombay and I was in India proper,

where almost nude men walked with their cattle or worked in the fields. They looked like the natives from the Mount Hagen region of the New Guinea highlands, with only a loincloth tightly draped about their hips, and bearing enormous loads of firewood on their heads.

Little creeks and rivers ran along squat hills, from where plunging waterfalls raced over grassy descents. The land looked rich and fallow, with the rain falling upon me with blueprint exactness. I was all plastic coverings, soon enough, covered from head to my jodhpurs until I reached Poona, where the rain stopped for an hour.

Rain, rain and more rain was my liquid umbrella through the lush Katraj Ghats, the soft billowing hills that ran between Poona and Khondapur. As I left Poona, I saw two cops hauling a man off to jail. His hands were tied with a long rope, which they used as reins. He shouted to the passers-by, who ignored his yelling pleas, and I wondered what effect this would have had in the United States, where prisoners are not hog-tied and dragged through the streets. But this spectacle became common enough soon enough, for I saw the same thing in Kolhapur as I came out of the mud and the rains. After the second police incident, I thought that I was seeing things, for I was constantly being blinded by the glare of the lights shining on the wet roads. Once I skidded in the mud trying to avoid hitting an ox. The driver merely drove on, though I couldn't, for the scooter was lying across my right leg and I kept slipping in the mud every time I tried to lift myself. But soon two peasants, who had observed me while sitting out the rain under a tree, came to my assistance.

Whenever I used the brakes, I skidded, cursing the monsoon that continued for hours. When I reached the Khambakti Ghats, the scenery became eerie, with the Indian mist dancing like apparitions on the intensely green, mossy hills. I bypassed Mahableswhar, for a sign said that the road was not open to motors, which probably meant scooters as well. Besides, I had been warned to stay off the road unless I was armed, my informant asking me in surprised tones, "What, you are traveling alone and without a gun? Are you mad, man? There are dacoits near Satara and you had better be off the road and in some safe hotel by four o'clock." It was a variation on an old theme, Iranian and Indian: "What, no military escort? You'd better be in Kolhapur and not around Atit by nightfall."

In Poona, once the military heart of Imperial England, I met an automobile mechanic who had given up a bad medical practice for a more lucrative mechanical one; and since Poona is the educational center for the Indian army, with every officer breaknecking down the slender roads on powerful motorcycles, the mechanic's change of profession made a great deal of sense.

"I am their amateur doctor for more intimate difficulties. If it's not a dirty spark plug or a clogged fuel line, then it's a dirty gonococcus."

The former seat of the Mahratta rulers, who had been defeated first by the Afghans, then by the English, still had its famous fort, though it had been turned into a Hindu temple. The remains of the Mahratta empire loomed up sentimentally, breaking into the shabby streets now filled with holy cattle, almost like a visitation from the seventeenth century. But instead of fiercely emblazoned warriors, military motorcyclists raced noisily by the Jain Temple and the Empress Gardens.

At breakfast I found myself beside an Indian Seventh Day Adventist, who offered up many dire warnings regarding the almost immediate end of the physical world and the beginning of a new world for mankind. "We've had World War I and World War II," he said over strong tea and crumpets, "which the Bible said we would have; and the atom bomb, too, so that everything has really happened just as the Bible said it would— so you must come to next Sunday's meeting. This is your last chance to square yourself. After all, you are a great adventurer. You've come all the way on your Vespa scooter, so why not venture a little bit more onto the road of glory?" The metaphor was quite telling and in keeping with his style of proselytizing.

There were many Gujarti guests at my hotel, the Dreamland, but I was not allowed to eat with them or in my room.

Not being a vegetarian, I might have contaminated their physical beings by eating meat. I ate my dinner off in a corner, somewhat ashamed of a sector of the human race and its insane symbolism; but when I protested the amount on the bill, which had, by some magic formula, doubled up in the peculiar service, the waiter remarked, "The proprietor has two minds and two bills for people like you, apparently," and the waiter tore the bill in half. "Now all things are evened up, sir."

After the dining room was empty and the floors scrubbed, the waiter put his bedroll on one of the tables, kicked off his sandals, and entered into Dreamland's noisy Nirvana; for most of the night there was a tumult and games, with the vegetarian Gujaratis playing hopscotch with the heavy furniture and the vegetarian furnishings of Dreamland's interiors.

The next day was a day of monkeys, monkeys and monkeys, dashing and leaping all over the road from Kolhapur to Sankeshwar. When I approached the dervishing monkeys, they were off with their babies, no more than a fistful of jabbering energy and movements as they jumped from the ground to the pipal trees. They swiped at the heads of the native women carrying heavy loads of kindling, or made madder antics at the almost nude men fixing the potholes in the road.

Outside of Karnagla, a small but typical Indian village of thatched huts, I saw a dead water buffalo going to the dogs and the huge birds of prey. The dogs encircled the dead buffalo, baring their fangs at me; and the birds of prey, a few feet away, jumped in and out of the ferocious circle, as they snatched bits and pieces of the carcass.

The Indians I talked with have a peculiar way of nodding yes and no. They reverse the affirmative and negative head movements, rolling their heads from side to side so that you never really know if no is no or yes is no. It's so easy to copy and offers up an adventure in dualism; and since I'd also become heir to Turkish and Iranian head movements, I found myself saying no and yes at the same time, which confused even me.

At the Chatprabha river, which is usually a raging torrent and overflows during the monsoon season, hordes of women were at the banks washing clothes, children, animals and brass pots. It looked like a ballet as they bent over, arching their firm breasts; and when they rose, they placed the now gleaming pots on their heads and walked off, their bodies swaying rhythmically and beautifully, much like the women who carry faggots and fodder, trooping endlessly down the roads from Pakistan to Ceylon.

At Dharwar, forty miles from Portuguese-held Goa, the small colonial enclave that Nehru will eventually attach to India, I found myself in a dilemma. Should I go to Goa to see the Tomb of St. Francis Xavier or should I go on? If I entered Goa, I would be forbidden re-entry into India; for that was the political payment one had to make. Did going to Goa mean one was sympathetic to Portugal, even if the mission was spiritual rather than political? I sat on my scooter trying to force this camel through my needling question. After many cigarettes, I decided not to see the tomb and ended my hour-long spiritual quandary.

My spiritual confusion was joined by a physical one at Haveri, when I had to give up my room at the Dak Hut or "rest house," which the Indian government rents cheaply to travelers, when F. K. Patil, the deputy minister of forests and fisheries, barged in with his retinue of civil servants. He had been at Ranibennur a few days before, and I had spent the following night there, inheriting his choice mosquitoes. I went on through the darkness and found another Dak Hut in the next town, experiencing this communal socialism at the ground level.

For the off-beat traveler, the Dak Hut is a thing of joy. The keeper, or *chokidar,* will cook for you and make you comfortable. You pay two rupees for the bed, eight annas for the water he draws for you, eight annas for the kerosense lamp, four annas for the sheets and the blankets; and whatever your dinner will cost, which he promptly goes to fetch from the nearest grocer and butcher. Should you use his toilet tissue, it will likely cost you so much per sheet; for the Indians love to keep accounts, giving you little pieces of paper everytime you make a purchase. It becomes a poetic flourish, a gesture out of romantic literature.

India is new each morning. There is the early mist on the

soft hills; a strange geography that evokes a sensual response. It is light and dark at the same time; and suddenly colors flare up, with attractive men and women making a luminous entrance. In Mysore, the changes in scenery became immediate, with coconut trees tenting over the soft landscape. The roads were always shaded, with huge trees of varied types acting like green umbrellas when the monsoon rains poured from the clouds. The natives were much darker; they wore earrings; and as I passed them their hands went up in supplication, or I imagined it was the baksheesh gesture; or, to the contrary, they may have been making a private little prayer for this mad Westerner on this strange vehicle, wearing dirty khakis, who appeared to be getting quite wet in the heavy noonday rain.

When, as it became my custom if I was too lazy to study my linen map, I asked how far it was from this place to that place, the answer always was, "Sir, it is ten furlongs." The word "furlongs" made me think twice, mathematically, for I always saw Saratoga Racetrack, off Yaddo, and some earlier mornings in August, 1940, when I was writing a book at Yaddo; and to break away from the intimacy of poetry, I objectified things by sneaking into the morning practice sessions at Saratoga, just over the hill and the woods, in back of Yaddo. A furlong, I visualized, was something many people lost their money on, by inches.

At a small café in Chitaldroog, birds flew in and out as I drank my tea and ate meat and vegetable cakes. The birds whistled over my head, and the audience that usually collected to see the scooter and the odd American laughed when I photographed them in their musical flight.

The hills were oddly shaped; more manmade than nature's own balances; more like totems piled on top of each other, with huge boulders encrusting them in wild, artful places. And since desolation always brought me back to the source, and was less taxing than our gadget-ridden scenery, I discovered that I was happier in the midst of it. Mount Ararat was like that, as well as the deserts and the mountains of Iran. The Indian hills appeared like sacred, shifting images changing every morning in some massive movement. I saw Calder, Miro, Buddha, and the hosts of the Lord creating this granite garden for my acute appreciation of spiritual and physical desolation.

When I reached the once-exotic city of Bangalore late that afternoon, I decided to go *pukka English,* to stay at a colonial vintage-type hotel, and give myself a good scrubbing. My clothes were sadly in need of laundering. I looked like a bindle stiff, scooter and all, much as if I had just dropped off a freight train. At the plush Central Hotel they looked twice, then thrice, before they had me sign the register with all their "verry" pukka accents ready to drop me in my tracks.

"What's this bloody thing?" asked the English proprietor, offering me a Scotch and soda.

"A Vespa scooter. I sit on it and it moves," and I moved in on the drink.

"But why not a car and comfort? Why not a plane and speed? Why not a ship and sanitation?"

"In an hour I'll be very sanitary, sir," I answered, laughing at his chagrin and my differences.

"I'll get the *chokidar* to get your laundry. Would you like a suit pressed?"

"Ah, how fastidious can we get?" And we had another drink on the house, with the accents slowly burning away, and the proprietor calling in to his very attractive blonde English wife to come and see this dirty American.

"Dirty, but in a rugged sort of way," she said. "You can't be an American—not a North American, I'm sure. Are you pioneering?"

"A New York American, which is quite in the north, or was, some months back. I have no American Indian blood in me, either." After that, pioneering came in for ten minutes of give-and-take banter, with the Scotch flowing away and the *chokidars* waiting to take all of my clothes, especially what I was wearing, off to the laundry. I finally managed to disengage myself from the mustached proprietor and his lovely wife, and off I went, heralded fore and aft by four servants in bare feet.

It became a vaudeville act as one took my shoes to shine; another one my suit; a third my underthings, and the fourth, who was supervising the operation, directing the etiquette of the drama. I was left completely nude, somewhat embarrassed, then told that my underthings would be washed and dried within a half-hour; that the suit would be pressed in fifteen minutes, and my shoes shined in five. All I had to do was take a hot bath and all would be back by the time I was finished.

All was back in time; and time was all I had, for I hardly ever looked at a clock. Time had evaporated like schedules; and only people mattered, or places and their people. I was never a traipsing tourist visiting temples and edifices; rather, I was seeing people at work or idling or being indifferent about their work and idling with a vengeance, approximating my own attitude from the non-restless scooter. I had seen the temple tourists; the fat, flatulent malices and heard the ejaculations of surprise that Indians, or any other people for that matter, had given so much to history. However, I reluctantly became a tourist when three Indian college boys on motorcycles became my unorthodox guides. They sped me, with tumult, clangor and horns, to the gentle Lal Bagh gardens, where I photographed the floral designs shaped into tigers, bears, elephants and other beasts of the jungle.

The three students were very kind; and over coffee one of them asked what one of his friends called "a rash statement," which was, to wit: "Why can't Indians come to America as tourists?" It was new to me, and I reacted with democratic aplomb, insisting that all an Indian need do was to go to the American consulate and get himself a visitor's visa, a point which his two friends accepted. This was followed up by the next stock reaction and question: "But why do they still hang Negroes in the South?" This stickler was answered in part by one of the other students, who said, angrily, "And we killed a million Pakistanis seven years ago, so how pure are we? They were almost the same color, too."

The next day they were my escorts and rode with me out of Bangalore as far as the police post at Hebgedi, where we exchanged addresses for future visits; after all, they had motorcycles, and they could tour America, as one of them promised to do when he left college.

I went from rain to rain until Toppur, where I watched the villagers weaving cloth along the road. The looms stretched down the road in seventy-five-foot lengths; and about them clustered numerous weavers following Gandhi's precepts about cottage industry; only the cottage was the highway and was much more industrious, yet with all the symbols of the anti-machine and anti-modern principals for all these heralds of anti-progress. The colors were not as interesting as those I had seen in Western India; it was as if some special image and coloration attested to the purity of their cottage industry precepts. I bought a piece of cloth for a few rupees and I was given tea and cakes as a bonus, with many questions being asked about the scooter, which also used a road but seemed to be associated with progress and speed.

At Toppur I met a surveying team, the forerunners of the new railroad that was to come from Bangalore to Salem—barefooted lads led by a scholarly, elderly, genial gentleman who was reading, of all authors, David Thoreau. With Thoreauesque concern, he gave me his extra cot, got me a private room, and looked fatherly whenever I asked questions, though he could do nothing about the non-Thoreauesque mosquitoes, which must have plagued Thoreau in another century and another country. But my sleeping and eating only cost me six cents, as I was enveloped by Walden, Indian style, in the largesse of the genial gentleman with the surveying rod.

I went through four little villages the next morning before I found one that had anything to eat, at least for a foreigner with his special stomach. And since something hot, like tea, is always safe, I had tea and stale bread and *gee,* which is not butter no matter how much it looks like butter.

It was a day that everybody changed his yearly string; for good Hindus, male and female alike, wear a string across their chests, signifying for the married women that they are happy and content with their husbands; for the men that they are at one with Buddha; for the children that they will grow up to be like their good parents.

Every village in the early morning has its wheat sellers, who empty a bushel on the ground, then spread it about in designs, over which they brood until a customer comes along. Trucks and buses ride right over the wheat, and I had to do the same, for the wheat sellers, when ambitious, spread it over the roads.

I passed a vaudeville act, Indian-village style; a man had trained parrots to climb up a small swing, to go through hoops, to fire a petite cannon, and just about ready for two weeks at the Palace, in New York. In the midst of the men and children viewers, I was soon the attraction, with my lost blondness and blue eyes something to stare at. It must have given them some sort of exotic pleasure.

The rice paddies gleamed in their sunken, muddy gardens of evening; and with more rain falling I was wet for a long time. When the paved road gave way to brown earth, near Vadipatti, I spilled, skidding into a bundle of fright. A passing bus driver and some of his passengers ran to help me; but an hour later, when it happened again, I knew for certain that wet, muddy, oozing roads are for mules, who do not have foot brakes or human reflexes that push down on them when going into a skid. The second time it was in the darkness and I couldn't tell whether I was going six miles an hour or sixty, though it was probably closer to seven.

I reached the elephant-panoplied city of Mandurai looking like an unwanted barefoot boy, rain-soaked, black, but feeling pale; as if the skidding had suddenly made me conscious of my so-far-fairly-free-from-accident trip. My reflexes were shot; and even the Hall of the Thouand Pillars and the Great Temple, with its nine gopuras (pyramid towers) was not photographed. I was subject to unconditioned reflexes, especially skid-reflexes; though, later, when crossing the wet, soaked, clay road across the Australian Nullarbor plain, I was to become a skidding expert, learning to take a fall as if it were a game I was playing with the skidding earth below me.

A festival was on, with richly decorated elephants plodding by, and bearers and musical heralds edging ahead, striking gongs and timeless graceful poses, as if the elephants were to be pleased before men, women and children. I stood off, camera conscious, fiddling with film, my dusty scooter looking somewhat bizarre alongside the decorous elephants and the enormous pageantry in the blazing sun. I joined the festival as the public photographer, with children and their parents, as well as the elephant bearers, trying to get into the photographs.

I was nearing the southern tip of India, a country I had seen too little of, bound for the sea frontier and the train at Dhanushkodi, which I would have to take to get to Ceylon, as the road went only as far as Mandapam, twenty-three miles from Dhanushkodi.

Mandapam itself was a small village, with a quarantine station run by the government of Ceylon, where Indians bound for work on the tea plantations of Ceylon were examined for communicable diseases and kept overnight in little huts; and if they communicated nothing too dangerous, they were allowed to proceed to Ceylon and some sort of a living, which they could not make in overcrowded, workless India. They got on the train at Mandapam, went to Dhanushkodi Pier and boarded a small steamer for the two-hour journey over the Gulf of Mannar. At Talaimannar, the Ceylonese Customs examined them carefully, looking for other things besides communicable diseases; for smuggling, as a great art, embraced many of the 750,000 Indians entering Ceylon every year for work or other forms of private enterprise.

It was a quiet night at the bare, clean quarantine camp at Mandapam, with four local Ceylonese officials paying a courtesy call on me. They knew the novelist Paul Bowles, whom I had met years ago when he was a composer. Bowles owned a little island eighty miles below Colombo, which Shaun Mandy, the editor of *The Illustrated Weekly of India,* had asked me to look over as a possible buy. The officials were interested in other things, however, especially things American and things of the body, especially the ways of love of white women. One of them, given to more philosophical detachment, thought that I could obtain perfect serenity by reading the epics and puranas of India; which would, in time, if read with real attachment, bring me to an even more poetic Nirvana. We had beer and tea on that, toasting to the excellent relationship between men and women, with or without the pursuit of Nirvana as an end-all or a cure-all for Eastern and Western man's contemporary difficulties. Another official gave me a copy of the puranas as a gift, and wrote the following inscription: "A memento of the fervid moments we passed at Mandapam Camp on August 22, 1956."

The next day I took the special train to Dhanushkodi, then a boat across the gentle Gulf of Mannar; a body of water stemming from the Arabian Sea on the west and the Bay of Bengal on the east. As I watched the slow liquid movement of the ship against the greening gulf, I was joined by a traveling salesman from North Carolina, selling rubber from Malaya, as well as anti-Negro tripe. I was soon very uncomfortable and said that I had written a sympathetic novel about the problem, though I hated the word "sympathy." It was not human enough; but I had worried about the reaction in the Negro press, which had given me better reviews than the white press. I mentioned it proudly, for whatever little pinprick it would bring. Yet he was a nice enough salesman, in his British tropical shorts, wondering if he could sell any rubber in Colombo.

Most of the passengers were poor Indians who had come to work on the tea plantations in Ceylon, looking spare and bare with their little bundles and their sleeping bags. Some got sick and made the deck a garbage can; while others contented themselves by looking vaguely at the sea as if that held a less sick fortune in store for them. Others were going to Colombo to take a ship for England, and eventual escape from India. I heard and saw quite a few Indians who preferred England and its certain democracy, its more certain employment, to the insecurity of Indian politics and work.

One poor Anglo-Indian had had his pockets picked on the train, losing his last hundred rupees, and he would arrive in London broke, without even cab fare to take him to his sister's home. We collected twenty rupees to help him out.

Chapter 8

ANOTHER EDEN

THE BOAT LANDED at 6:40 P.M at Talaimannar, but it took the Ceylon Customs three hours to clear me through; and since there was no place to sleep, I scooted off at 10:00 P.M. for Mannar, twenty miles away, where there was a rest house. I was put up in the dining room, for every room was occupied; and I slept amid cups and saucers and drinkers drinking late beer, wondering what Ceylon looked like during the daytime. I saw elephants, tigers, tea plantations and all sorts of wild game; and I saw myself in an accident, so that my sleep was all nerves. When I got up at six, I was worn out with tiger hunting and the fear that an accident was upon me.

I planned to take two days to reach Colombo, about two hundred miles from Mannar; to take it easy and use my camera. After all, there were elephants, tigers and my fears. I left at 8:00 A.M. and headed for the temple city of Anuradhapura, about fifty miles away, hoping to make it before lunch so that I could have the early afternoon at the great white temple, Ruanveliseya, as well as the sacred Bo-tree, where Buddha had meditated 2,500 years ago This tree, still growing, had reforested the spiritual essence of India's living trees, for branches from the original tree had been planted all over India. And I rode carefully, philosophically, with the wonderfully cool breezes off the Arabian Sea blowing ahead of me. I saw hordes of monkeys on banyan trees, but they scampered away before I could reach them with my camera. I made four attempts, with one baby monkey allowing me within ten feet before it screeched off to the highest limb.

Ahead of me was a bus filled with teen-age schoolboys watching my futile monkeyshines as I tried to get the monkeys to pose like human beings. After each photographic failure, I trailed carefully behind the bus, seeking out more chattering trees and their agile populations. Twenty minutes and two miles later, wearying of this play, I blew my horn to pass the bus. But the bus, which should have been on the left, as in England and in India, was over to the right, and no amount of horning at the driver could get him to his rightful side, which was on the left. I had, by trial and without too much error, in India, gotten my reflexes accommodated to the left side; but like the monkeys, the driver was not accommodating me on either side of this very narrow road.

I pulled up, yelling to the schoolboys that I intended to pass on the left, since the driver was not giving me my rightful if innocent passage on his right. When one of the boys called back for me to go ahead, I moved up slowly, hugging a few feet behind, thinking that my intentions were understood by the driver as well as by the acknowledging schoolboy. Starting to pass, I gave the scooter the gas. The bus swerved, hitting the scooter and rolling me, at fifty miles an hour, onto the road.

At Mandapam Camp I had been given an anti-cholera injection so that my right arm felt sore and I was still slightly feverish, twenty-four hours later. My responses were inaccurate and my judgment faulty, or I would not have tried to pass in that fashion. And as I lay on the road, my chest felt shattered, for I had been thrown upward, falling chest down. My head was bleeding, with blood oozing into my gasping mouth. I couldn't breathe. I tried calling to the boys on the bus that I was dying. It felt like death, in a breathless world, my chest unable to heave in or out.

The scooter was still running, the wheels jabbing into the

dust, aimless and peculiar. I expected it to explode as I was exploding in my lungs, unable to find air on the ground. Then a sound finally broke from me, a great echo of terror saluting the howling monkeys in the nearby banyan trees. The boys were now running toward me. I raised myself a few inches, to fill my lungs, to turn off the motor, and I fell back, thinking I was unconscious or dying again, but I was only too aware that somebody was applying smelling salts, that there was blood on the ground, that my clothes were bloody, my knees a mass of blood breaking through my torn pants . . . that this might be the end of my self-sought journey around the world or to nowhere . . . "The thing which I greatly feared is come upon me."

But an hour later my journey to everywhere was renewed. I was a bandaged passenger riding on the almost-fatal bus, my scooter nestling awkwardly on top, having been raised there by the boys. My knees and forehead were badly bruised; my chest felt as if a hot ramrod had gone through it; but no bones were broken, so I was assured. I was merely a much frightened man, I was told in dulcet tones. Did I intend to sue?

No, I did not intend to sue, I told the manager of the touring boys, a Catholic dignitary who had been showing his charges around Ceylon for a week. No, I wouldn't sue. "Just get me to a doctor and the scooter to a Vespa agency and let all things be repaired and on their way."

It took an hour to reach Anuradhapura, where the doctor was, and from where the scooter, much damaged, would be sent on as dead freight to Colombo. But this was my day for accidents. As the driver neared the Anuradhapura railroad station, he ran the bus underneath a tree with low-lying hanging branches, and the scooter came hurtling down, swept away from its precarious perch. The police, who had been following the bus to the station, where I was to give them a report of the sideswiping accident, looked on with amazement when they heard the crash. Now everything seemed insane at once, and I was assured by the boy next to me that the driver was an escaped madman running over strangers, scooters and chattering monkeys, which he had done the day before; or that he was anti-Point Four, with peculiar ways of expressing his anti-American antipathy. As I stared painfully at my now much-wrecked scooter, I said to the manager and the driver, "Sorry, but I think that I'll have to sue a little bit, if only to get this madman's permit to drive revoked."

The police acted like police, with pens and paper much in evidence. I made a statement; I had been photographing monkeys, etc. The driver made a statement, etc. Everything was now very official, including my bruises. A week later, when the repair bill came, the bus company paid more than five hundred rupees or one hundred twenty-five dollars for the driver's strange exercises in political disenchantment.

The doctor's examination and X-ray showed many bruises but no broken bones. Joyous and unbowed, I met two law students at the hotel where I was resting who were driving to Colombo. En route I was shown around Anuradhapura; and feeling strangely holy, I made obeisance to all things holy. I had not been killed. I bowed at the sacred Bo-tree and kissed the earth, touching the tree that had started its pilgrimage on earth 2,500 years ago. I bowed reverently to the monoliths of the former Brazen Palace, feeling like an angel of life come home, via scooter, for another look upon earth and man.

At the huge white shrine, the Ruanveliseya, with its columns of gray painted elephants, I bowed again, feeling how meek is the voyager indeed and how weak are his bones, when colliding with spiritless things. When I saw the statue of Buddha, I was almost ready for Buddhism, in a charged conversion, with my instincts telling me that I must have been spared for a worse future, so I had better be at one with Buddha right now. But my eyes could not focus and my brain could not respond. I was numbed and in shock, shaking as I walked about the moat cutting off the people from the shrine. One of the law students had given me a tranquilizer, which was new to me and likely old to him, and it helped make the journey to Colombo tranquil. I fell asleep, dreaming that I had been in a serious accident; that I had foreseen it; that my sleep was only escape from the real . . . sprawling on the road and dying. Five hours later, we reached Colombo and a Chinese hotel, where I slept for twenty-four hours in a flight of many fancies.

The Ceylonese poet Tambimuttu, who lives in New York, had suggested that I call in to see his relative, Jayanta Padmanaba, a former editor of the *Ceylon Daily News,* but now a columnist on the paper. When I met the handsome, affable Padmanaba, and he saw my array of bandages, he had a staffman, Oscar Rajasooriga, do a front page story: *American Writer Here on a Scooter, Almost Killed by Madman*—with a photo, which one of the boys on the bus had taken, showing the scooter being raised on the top of the bus. It was an excellent story from many angles, for the insurance company paid up; and wherever I went in Colombo or in Ceylon, I was a guest, with many apologies for my pains. It was not customary to run down Americans on scooters, people said, especially when they were photographing Ceylon's beautiful monkeys.

But though Ceylon is a photographer's paradise, it is not so for the journalist who wants more than tropical scandals for his typewriter. After a few days I understood why Frank Moraes had gone to Bombay, and why Tambimuttu, who acts like a prince in New York, had fled from Ceylon. There is nothing to cultivate for a "prince" or for a hardworking journalist after working hours. There is no dance, which India has in abundance, though there are the Kandyan·dancers at the right season. One can hunt tigers, however, or drink himself under rattan tables at Latvinia Beach or at the Galle Face Hotel. One can go native, with little reward in this Eden that is balmy and palmy without goddesses. For Ceylon is a political garden growing a variety of Marxism, with the contemporary Trotskyists blooming strangely within the government, which they do not oppose, as a loyal opposition.

But since all things are functional in Ceylon, all Ceylonese function after a fashion. There are the Dutch Burger girls, white in part, exotically, racially mixed. As for art, it is the movies, or second-rate imports—bad singers who do badly in Europe and tour the faraway places with much reward.

Ceylon is a tremendously beautiful wasteland of the mind,

with nature offering up astonishing Edenesque substitutions and enormous natural pageantry, as well as tigers, hill stations, political scandals, and many driving accidents, or petty criminality, which of course is hardly objectionable in Eden, with its ancient temptations. Colombo, which is a port of call en route to more vital places, suffers from a hallowed dullness that even lush nature cannot replace.

This was not my complaint at all; for I was happy despite my wounds. I sat around with the journalists and listened to their badinage. "I want to go to Europe or to Australia," said A., the young journalist of the *Daily News*.

"I know some Burger girls, but you have to woo them," said B., the staff photographer.

"See Ceylon and drop dead from boredom," said C., a half-Burger, who specialized in writing crime stories for home consumption and travel articles for the New York *Times*.

"I'm going back to Singapore," said tall, excitable D., a sports writer. "We can always chase communists there and call it sport, if not politics. There are night clubs, Eurasian whores, concerts, and more communists. They have wrestling with unmasked marvels, and games that are not so taxing, as well as more rubber communists, who bounce back every time they are reported killed. What a place!"

"I'm a movie expert," said C., "made so by nothing to do but go to the movies every night. I've now been hired on a part-time basis by Sam Spiegel, who's doing a new picture here. He's going to have a real bridge blown up, and all the religious organizations are protesting, saying that this is an exposition of violence and a bad example for the children of the country. Imagine that!"

Said the poet, E., "What shall I do with my poems here? I am out of the times; for no one, including my friends," and he pointed to the other journalists, "reads poetry here. Shall I send them to a publisher in New York? Do you know a publisher who will publish me? No one will publish anything except invoices here."

It sounded much too brutal, even as banter over beer;

and A. was saying, as if to apologize, "It's the tooth of Buddha that has everybody peculiar here. They have the Temple of the Tooth in Kandy, and that's supposed to make everybody gnaw at his soul. One tooth of the Lord Buddha, and nothing else."

"What has Buddha to do with the twentieth century?" asked C. "The religions of the world need their cavities filled, or just plain new dentures stuck into the mouths."

Cynicism and wisecracks did not make the Ceylon garden blossom any better. As a matter of fact, there was even a shortage of water, with much fear about the future crops. The tropical climate had robbed the alert Ceylonese of his mental excitements, giving him much in nature, though he could not escape from his natural boredom.

"But there are fourteen Trotskyists in the House of Representatives," said C., "and we also have the hero of of Bandung, Sir John Kotelawala, whom you must meet, Mr. Harry."

All the journalists had Vespa scooters, which clustered in front of their offices at the Fort, making it all very sporting to look at as they dashed off to cover accidents or the general mayhem that made vital statistics an impressive art in Ceylon.

"Get your scooter back from the shop, Harry, and we'll all scooter up to Kandy on Sunday," said A., who went off to call the repair shop. When he returned, he said, "It's going to cost that bloody bus company a lot of rupees, Harry. It's being repainted now, all ready for our trip to Kandy."

"Bloody lucky you were not killed," said C., and he quoted facts and figures on Colombo's accident rate. "Two a day are killed on Galle Road and Marine Drive, so we must have the most reckless drivers after Paris."

"Also a new type of call girl," added D. "They use cruising taxis with their picked-up clients. But if you want a more primitive kick, you can have a Tamil girl in a bullock cart. All you need do is to go to Lipton Circus and soon enough some pimp will accost you, take you to a waiting bullock cart, which you enter. Inside is a naked girl waiting for a customer; and off the bullocks trot to Colpetty. Call girls without telephones. Quite primitive here, aren't we?"

I went to Kandy on Sunday, but without the four frantic journalists who found, upon Buddhist reflection, the whole idea much too exciting for their nonwriting day of rest. It was easier to complain. And besides, they even hated nature, or getting out into the "bush," as B. said; he had done a Colombo Plan study stint in Australia.

I was in elephant country within an hour, suddenly "away from it all," enveloped in beautiful patterns of tropical, natural violence. Along the narrow road, with the usual mad drivers tumbling about in Volkswagons, I saw one turn over, then right itself. The driver went on, his head intact if his fenders somewhat bent. Along with the mad drivers, for it was Sunday and sunshine, the road had lovely girls selling *cadju* nuts. The girls wore colorful saris, their hair adorned with red and blue flowers that made a kind of burning halo about them. At their tables they split the *cadju* nuts with a long knife, offered the coconut milk to the buyer, then tossed the shattered nut to the side. On Monday a truck would make the rounds and the nuts would be processed for soap and oil.

Prime Minister Bandaranaike had his country home in the area, with a typical Ceylonese garden facing the road. In the middle was a small pond, where the Minister contemplated on weekends as he escaped the noisy and insulting quarrels of Ceylon's many waspish politicians. A hundred yards from the house, a girl was working in a paddy, her skirt lifted high above her knees, showing her round, firm thighs; and it must have been like this when the Garden of Eden bloomed, if without the ministerial associations. The girl's strong hands moved delicately but precisely as she bent to the earth, her body making a sensual, living arc. Her feet were sunk in the oozing paddy, yet her skirt was completely clean. When she stood straight, her up-jutting breasts were lilting and conical. I went on.

I saw my first lumbering elephant, a baby one lying in a small pool of water. A naked boy was scrubbing the elephant's back, with the elephant disobeying the boy's command that it turn over and get its belly washed. Instead, it waddled up and came down on its haunches, with the exasperated boy yelling commands, insisting that it come down on its left side, which it did, after more elephantine baby-playing.

Across from the pool was a large plantation of coconut trees, making a soaring, leafy foreground for the green laden mountain rising behind. At the foot of the mountains were more elephants; large, working beasts in steel chains and harness. A dozen men were leading the elephants up the mountain to root up coconut trees that had died; the elephants curled the trees within the embrace of their trunks, much like baggage being carried by porters. They went up and down the mountain, obeying every command of their boys, their chains rattling fiercely as they scraped against foliage, rocks and trees, scooping up the surface of the mountain, much as if a cyclone had suddenly swooped down and erased everything that grew tall on the mountain.

A week before, Kandy had observed its annual *Perahera* or procession, and the Temple of the Tooth put the tooth of Buddha on display. The Dalada Maligawa is the soul of Kandy; the tooth, which is seldom exhibited, is its spiritual force; for after sixteen centuries it has taken a religious bite into time. A silver bell-shaped shrine is its protector, along with six more inner shrines, all laden with precious stones. Once Kandian kings had fought over the tooth, moving it from capital to capital; but now it is permanently housed, not too far from neon-lit restaurants, art and craft shops and electrified gardens, all without Kandyan dancers to add a balance to contemporary garishness.

"All the tea in Ceylon will not remake the place," A. had said over the phone that morning, warning me that I was expecting too much too soon, "There's nothing there, really—nothing at all, old chap." But there was and always is something for Americans haunted by a lack of ancient sculpture, by time's brittle conceits regarding antiques, art and ancient man.

I saw the Kadugannawa, or the Bible Rock, much more like an apparition invading a valley, its flat top nestling over the steaming valley that was soon raining geyser-like, with the monsoon pouring upon me. I left the road and found a glen not too far off, with shade trees as umbrellas, my world suddenly a watery garden. Thick foliage ranged about the secluded area, with the down-thrusting rain beating a massive tattoo of sound. It was Eden, but wet.

I ate my sandwiches and lapped at the rain, somewhat disconsolate and yet awed by the lonely scene. I heard a car swishing down the road, just over the rise, for I was less than a hundred yards away, in the semi-jungle that traced itself beyond the trees.

When the rain stopped, I started back for the road, my scooter kicking up spray and wet sand; and as I moved onto the paved road, a passing car hurtled by. I waited, looking at the lush greenery, the tall coconut palms rising above nearby native huts; then more screeching brakes as the Sunday drivers from Colombo sped back from their visit to Kandy.

My scooter was soon going forty, but I slowed down to twenty-five to make a turn, for the road twisted up and down in looping spirals. As I did, something rushed from the side of the road. My instincts were automatic, knowing that fear is a first cousin to bravery or cowardice. When I looked, I leaped, or my scooter did, racing around the turn at its widest arc, shooting down the straightway that ran for several hundred yards; I raced at sixty miles an hour, with the down-grade helping me to get up speed—for the first tiger that I had ever seen in my life was behind me. I made each successive turn like a racing driver, no longer the prudent and careful traveler. I had experienced enough lacerations and bruises in Ceylon without the mauling a tiger would contribute. I was a speed demon burning brightly in the late afternoon.

I heard a gun go off several times. A car sped by, with the driver making crazy gestures for me to stop. For an insane moment I thought this was a plot to get the tiger some fresh Western poet, not too thin, but juicy, for a snack. When the car cut across me, I braked, skidded, then fell, but without any damage. I had my prudence back again.

"The tiger's dead," said the driver, who turned out to be a lawyer. "I saw him take after you. Do you want a trophy?"

He was holding a hunting rifle smelling fresh from used bullets.

"No, I just want to stay alive until I leave Ceylon," I said, somewhat nervous from it all.

"Then take a photo of me and the dead tiger," said the vexed lawyer.

I shook my head sadly, saying, "Today I did not bring my camera. You see, my camera had an accident," and I told the lawyer about the sideswiping incident.

"I read about it," he said with a loud laugh. "You must be slightly crazy not to have a gun or an armored car when traveling here."

I agreed, I must; but this was Eden, where one only had to look after snakes or friends of snakes and not bus drivers, who were not friends of anybody, though perhaps they had snakes for friends.

"Are there any more tigers here?" I asked timidly. "Would you care to escort me a bit? Was the one you shot very big?"

"Not very, but capable, I'm sure. If you'll come home I'll show you a few trophies."

I followed him back to have a look at the tiger. It was big enough and I didn't care to measure it dead or alive. The lawyer pulled it off the road, so as not to frighten other drivers careening back from Kandy.

I had an escort; rather, he rode behind me, to fend off wild beasts. In his home near Mount Lavinia he became the exuberant, sporting host, for he hunted, fished, and practiced law at one and the same time. His study had as many tiger skins as books, along with stuffed snakes, which seemed strange for Eden. But I had gotten used to the foibles and folklore of the Garden, especially when drinks were served, for Adam would have been out on his feet before Eve got around to offer him an apple.

A few days later, the *Ceylon Daily News* photographer, Wally Perera, took me at 6:00 A.M. to meet Sir John Kotelawala, the ex-prime-minister. Sir John was another explosive man, with a great deal of natural charm for all but his enemies, and

I gathered he had many. He looked upon politics as a many-sided game to use up his prodigious energy, which he leveled against communism as well as colonialism.

He had also, on various occasions, or so he told me over tea, called Krishna Menon, Nehru's international plenipotentiary, a political rascal, a fool, a loudmouth and a braggart; and that he was *persona non grata* in Ceylon while he was prime minister. At the famous Bandung Conference of 1955, Sir John had come out against all forms of colonialism, pointing up Moscow's variation on this contemporary theme as a case in point.

At the age of fifty-nine, after thirty years in Ceylon's overheated politics, he had become the prime minister, a post that he held for only two years. He was badly defeated along with his party in 1956 for refusing, he insisted, to promise the electorate everything under the Ceylonese sun by tomorrow morning or the day after, at the latest.

"But what is one to expect here, especially when there is no tradition about democracy? Here we are all busybodies, with every housewife wanting to know what's going on in the next apartment. In England, where they've had democracy for hundreds of years, people are scarcely on nodding terms with their neighbors, even though they've lived in the same building for thirty years. When they meet in the London tubes, they manage a polite enough "hello" and then take up their newspapers, ignoring each other completely. I suppose that's how you keep a democracy democratic and not have it fishwived into gossipy smithereens."

Sir John lived on a large estate on the outskirts of Colombo; the front of his huge house faced the airport, on land which he had sold them. He operated a large plantation of varying crops, using his elephants to till the fields, and I saw two of them pulling hoes. Wally Perera had published a photo of the elephants laboring in Sir John's many vineyards; but it was strange seeing them at work against the green hills beyond.

At six each morning, Sir John held an animal parade, with every animal on his estate doing a march past. He was dressed in traditional Kandyan robes, with vest, sash, spangled shoes and a princely hat; and as the animals of the Lord traipsed by, Sir John gave each a piece of candy suitable to its animal tastes, size and weight.

"It's odd," he continued, returning to politics. "Communism in fifty years will not be recognizable at all. It is going toward democracy, and democracy is going toward communism —and they will pass each other en route, taking from each according to their needs."

When I asked, "After Nehru, what then?" Sir John threw up his hands in feigned horror and shouted, "There is only one man in India holding the country together or keeping Moslems and Hindus from murdering each other. But countries learn to manage without their great men; for even Ceylon does without me. Someone will come up to replace Nehru. But God knows who it will be!"

His political party, the United National Party, had lost fifty-six of its sixty-four seats and was reduced to eight—a hopeless minority and hardly a formidable opposition, Sir John notwithstanding. The largest opposition, the Trotskyists, supported the government party, but they would have hardly been recognized by Lev Davidovitch Trotsky.

In his green sport shirt and white trunks, Sir John was indeed colorful, with his vibrant personality and his language putting him into sharp focus. Exaggerations went with his flamboyance. On the lawn were elephants and riding horses, and he waved to them like a father.

"All of this is going soon, for I intend to give these 105 acres to an orphanage," he said, though he was still questioning which orphanage would inherit it. "I might even give part of it to you and Wally," he said to us with a laugh; "then you can ride herd and go in for politics. I also have a hundred-acre estate in England, and the government threatens to take over all the unoccupied houses on the estate. But I will fool them, for I intend to transport to England six Ceylonese families and house them in the six empty houses. Why have them stew here when they can stew much better in England, rent free, for Sir John?" His loud laugh rang over the fields. "This is the age of private socialism, after all, isn't it?"

After tea and cakes he went back to the stewing problem, or international stewing, as he called the United Nations. "We must remain neutral in Ceylon, for we are too small to be anything else and can only stew in our antique juices here. We have no choice; for we can be swallowed up as we have been over the centuries. First came the Portuguese; then the Dutch; then the English—and no doubt it will be India next time. When my uncle was premier, we kept the Russians from infiltrating; but now, God help us! The Russians never play according to the rules; but we still do, unfortunately. Then, too, Ceylon needs a lot of help. We import a great deal as our land is given over to better paying crops, like tea; but you can't eat tea leaves. As for the Suez controversy, I am sure it will quiet down and everybody will save face, if nothing else."

Carrying a copy of his book, *An Asian Prime Minister's Story,* I was escorted out by Sir John, who introduced me to a few of the trumpeting elephants in passing. This was another Eden, with its inlaid exoticism, its animals, its gardens, and its gods. I had seen two Gardens of Eden; one claimed by Turkey for Mount Ararat, under the eternal ice of the peak's solitude; another in Ceylon, where the climate, if not man's own tropical sophistication, kept man from eating the universal apple of life. But this was Ceylon, with its open, vivid, radiant pageantry; a country of temples to the Lord Buddha, and where I had almost died.

The S.S. Stratheden took me to Australia, seven beautiful sailing days on the Indian Ocean. We passed the Seychelles, where Archbishop Makarios of Cyprus was having an enforced holiday, with England paying the lodging bill and calling the tune for Cyprus. But, then, England was protecting the Turks, who were a minority in Cyprus; for the Greeks, who were a majority, had all sorts of memories of 1922 and Ismir to incorporate into their sudden anti-British music.

Chapter 9

"TERRA AUSTRALIA"

WHEN I left the ship at Freemantle, fifteen miles from Perth, I was to stay with the short story writer Gavin Casey, whom I had known when he was head of the Australian Government Information Bureau in New York. Knowing Gavin's Australian love for things in bottles, I instinctively hid my short supply in my baggage, aware that before long, en route across the Nullarbor desert, it would be my medicine to fit every situation.

When my rear tire blew as I headed for Perth, I almost lost my liquid magic; for my baggage bounced off when the scooter took me for a header. I had gone almost 16,000 miles without changing a tire or adding air to them, which is a record of some sort for something of a sort.

Isolation without population; that was Australia in 1947, when I last visited the continent below, isolated by the flight of its own talents, which Henry Handel Richardson, the great writer, had begun. I had seen so many of my friends do the same. Sidney Nolan, one of Australia's most original artists, had fled to Italy and Greece; and I knew a hundred Australian men and women, living in Paris, London and New York, who felt a sense of frustration with the cultural climate at home. Nolan had once illustrated a book of my poems, and we had a kinship, for we had both worked on the famously hoaxed *Angry Pen-*

guins magazine, edited by Max Harris in Melbourne and Adelaide. I was the American editor, getting writers like Dr. Sidney Hook, James T. Farrell, Henry Treece, Dylan Thomas, Sir Herbert Read, to add an English and American blow-torch to Harris' literary exercises.

But the country had changed in nine years; for the seven million population of 1947 had grown to almost ten million, with half a million English and another half million Middle European migrants adding the excitable flavors of their countries. Australia, still very much in transition, was now showing some remarkable maturity in arts, politics, architecture and business. From the dock at Freemantle one saw the new and the old in juxtaposition; much as if I, a New Yorker of vintage 1957, were suddenly allowed to have a look at the primitive industrial specter that was New York in 1900.

But the images collided, exploded and reshuffled—for Australia was coming into a dangerous age, with its old White Australia policy being buffeted about by the nearby Orient, now only too conscious of the racial implication involved as well as the extreme isolation of the white population on the Australian heartland. Population from Europe and the United Kingdom might stave off or postpone the coming of the unwanted migrants from Indonesia and China.

Gavin Casey, who had run the Australian Ministry of Information at Rockefeller Center like a bohemian sporting club, was now a columnist on *The West Australian,* Perth's largest newspaper. He was a crusader, chiding the readers, the owners of pubs, and all the citizens for being satisfied with slummy-looking pubs and hotels, for Australia's bad cooking, for negative interest in everything but beer. He lumped them all in one pot and said they had but one interest in life—to keep the glass filled at all hours. Casey, in true Australian style, attacked Australian deficiencies in a comic fashion, keeping his own glass filled between chores of pounding out essays on Australian frailty in matters universal.

After three days, I was ready to go bush, to scoot across the grim Nullarbor desert and plain, which began after Norseman, 450 miles from Perth, and ran for over a thousand miles across the most desolate country in the world. It could take in the vast overpopulations of Indonesia, providing there was water to make the desert bloom. But it was barren and sandy, filled with saltbush and mallee scrub, unable to feed even the few aborigines living there . . . and I was going to scooter over it toward Adelaide, to see Max Harris, to give lectures at the University and for the Fellowship of Australian Writers, or so Max had written me while I was still in Colombo.

Filled with Casey's good beer and cheer, I left Perth in the rain and the chill, with Gavin saying, "Call in and see my brother Allan, who works as an electrician at the Kalgoorlie Brewery. Give him my best; and when you see James T. Farrell in New York, remind him that he's stopped drinking." Farrell had come through a few months back, lecturing for the Australian Committee For Cultural Freedom, with Casey's cheer and beer augmenting Farrell's whimsical detachment about things nonpolitical.

A bitumen road ran to the frontier town of Southern Cross, which I reached late at night. Overhead, enormous stars studded the dark blue sky. I remembered Hart Crane and a reference he had once made to the Southern Cross in a poem of intense, almost hallucinating phrasing. He later committed suicide. But Southern Cross, as a town, was alive with the sons and daughters of pioneers who had settled it a hundred years ago. This was the frontier and the outback; and I saw much of my American past riding herd along the lonely wheat and sheep country, which thinned out after Kalgoorlie, where the famous Golden Mile gold mines had been discovered in 1902, in a setting similar to Sutter's strike a half century earlier.

I went gold mine prospecting with Edwards, the manager of the Coolgardie mine, as a preliminary to visiting Kalgoorlie's Golden Mile. Edwards took me down the elevator shaft to rock bottom, 1,500 feet, after which we crawled through dripping granite caverns, hitting our helmeted heads against golden encrustations jutting out from all angles. A cathedral of death dripped from my fancies; and though I felt the chill and the panic of sudden enclosure, much like being entombed, I was not overfrightened. I heard the drills going in various chambers and the digging tools and hammers knocking against the dead-end walls. Horizontal and vertical structures laced the myriad caverns, which, when emptied of gold-bearing ore, were left to their past silences, except for the man-made buttressing frames that still decorated their dripping, deadly depths.

Later that afternoon, I scootered to nearby Kalgoorlie's famous Golden Mile's Croesus Mine, to watch tons of gold bearing ore, some of it from Coolgardie, being refined, and I got pay dirt, a few pounds of ore as a gift, with a dollar's worth of gold promised should I care to process it. I went down the 5,000 foot depth in a series of shafts, into the cold tunnels that gave man his basic currency and his molten fevers. This was the Golden Mile, an incision into Terra Australis, that had once given Kalgoorlie almost 45,000 people, who lived their nongolden days within the chambers of this precious metal.

I stopped at Kalgoorlie to see Gavin Casey's brother, Allan, for a ten-minute beer at his brewery. Ten hours later, I found myself chattering and sodden, and I loathe beer; but Allan's cobbers (Australianese for "friends"), who first put me on an electrified hot seat to test my humor, insisted that to scoot across the Nullarbor it was best to be slightly unconscious, "and you need fattening up for the hungry, beerless, waterless ride across that bloody sand where even aborigines bloody well fear to tread."

The solemn Eyre Highway began at Norseman; a clay soggy track across the Nullarbor wasteland, good for covered wagons but not for the small wheels of a scooter, which only increased the hazards and dangers jutting up every foot of the way. The road was built in 1941, when war made transportation from Western Australia to the eastern cities a military necessity, with sudden death from Japan threatening Australia's isolation. But sudden death shot up from the sudden softness, from the potholed maddened track that tossed me up and down like a roller coaster going berserk. I discovered the foot brake, which kept me from committing unwillful suicide as I catapulted from pot to hole and back again.

At Norseman, where I stayed for the night before entering the wasteland of the beautiful sounding Eyre Highway country, I was warned I'd find the occasional water tanks usually empty; that every two hundred miles there was a pioneering homesteader, who usually carried bacon and eggs as well as petrol, but that I had better pack everything for the thousand miles; that men had died of hunger, thirst and exhaustion when the heat hit, for it was, in the wrong season, as hot as Iran (which did not help me psychologically); that accidents were common, especially when running into the many sand banks on the Eyre; and so I packed the lot, with a grocer handily suggesting tins of this and that, along with a five-gallon water bag that I hoisted over my front baggage carrier. I was a new sort of pioneer, with small wheels taking me across the plains, where horses and nonsodden men feared to go.

At the gas station in Norseman, I was told by the owner that he had been on the lookout for an American on a scooter going round the world; that he was happy to give me my full tank capacity for free, which was a gallon and a half, providing I stayed over and lectured to his Rotary Club on places strange —"So will you please give a talk on the state of the world you've passed through."

I talked into his tape recorder, right then, with the owner shooting entangling political questions, jazzed up for a lively one-hour Rotary Club luncheon, with the tape as my political proxy, and off I went, suddenly adrift on the sand and clay that sprang into reality as I scootered out of Norseman toward Balladonia, 138 miles into the wasteland.

The saltbush was everywhere in the flat, dreary plain I faced; the road was a mass of jagged, scooter-halting tangles of natural disorder. Between Norseman and Balladonia there was nothing and nobody; but once I reached Balladonia, which sang so sweetly, I would find a homestead, with a Mr. Jackson selling gas as well as steak and eggs, providing he had something to sell.

An old telegraph line ran along the road, breaking into this primitive outback with a sudden touch of civilization. At a place called Buldania there was a water tank, but without water. It was also dry at Bedonia. These places were just names without people; they were liquid fancies of the Australian imagination, strung along the map to signify the human touch, tanks with names scratched on their sides, signifying nothing for the traveler entering this eerie, improbable and impassable land.

It was touch and go for many wearying hours, with no letup. When I stopped to make tea, I saw kangaroos dash across the salt flats, bounding away in fear. Now and then a colorful bird perched on the sagging telegraph wire, singing some bird song to me. I was alone for eight hours, though I had been assured that a half a dozen assorted vehicles were bound to pass by. If I had a breakdown or accident, I would get help, eventually. I kept moving if I shifted constantly from first to second, bumping and hurtling through potholes as big as a horse, averaging less than ten miles an hour.

Late in the afternoon I saw a station wagon bounding toward me. It was coming over a rise, careening down a sandy drift. It went from side to side, turned a somersault, went up and over again, then righted itself, much like a tragic circus performance. The three passengers, two women and one man, finally emerged from the station wagon. They looked stunned and in pain as they staggered about. I ran for my bottle of cognac, to help revive them.

An hour later, the three were still very shaky. My bottle of cognac was half gone—for we had tasted and toasted to no bones broken. But one of the women, an American shell-collector, complained that she had a headache. Her eyes looked sunken, as if she were suffering from a concussion. The others did not say anything, fearing they would frighten her with talk about concussions. I offered to go back to Norseman with them, to leave my scooter in the bush; but the man insisted that everything would be all right, saying, as they left, "It's taken you ten hours to get here, and you may not get a lift back. Thanks for the cognac, Yank. We'll be all right, chum."

A few days later I heard through the bush wireless that the woman had been hospitalized for a concussion; that the two other passengers were badly bruised and also in the hospital.

At dark, I was still many sandy miles from Balladonia. I saw a campfire, with four aborigines standing about. I stopped, made tea, and we had a communal tea party. The aborigines looked at the scooter with much awe in their dark eyes; and when they touched it, it was as if they were touching a magical vehicle.

When one of them said that they were from the Aranda tribe of Central Australia and had been wool shearing around the Bight, I asked the oldest of the men, "Do you know a song called 'The Song of Ankotarinja'?"

He began to jabber the song, breaking into the jagged, wild movement of the Aranda fertility dance not often done for whites at aborigine corroborees—the frenzied, frightening dances they stage when they want rain. His three companions joined in, and I went along for an exercise in primitive rituals, hoping it would not bring rain, which would bog down my scooter. As for the song, it had been translated by the Australian anthropologist T. G. H. Strehlow for an anthology of Australian poetry I had edited during the war.

The aborigine version and its translation follow:

> Nomabaue rerlanopai
> Nomajatin tjelanopai
>
> Nomabue relanopai
> Nomaalba tinjanopai
>
> Nomabaue rerlanopai
> Nomatnjenja lbelanopai
>
> Nomaarkwe rkarlanopai
> Nomatnjenja lbelanopai
>
> Nomakante kantanopai
> Nomatnjenja lbelanopai
>
> Tnimaruburuba laitnibe
> Tnimawurubina laitnibe
>
> Tnimanatan tjalitnibe
> Tnimawurubina laitnibe

Ratuwatela lurbmaturatu
Ralilertjala lurbmaturatu

Watuwatela lumbatnuwatnu
Walilertjela lumbatnuwatnu

Limanowo bintjintile
Limaldule ratjintile

Limankule ralintile
Limaldule ratjintile

Maraminjutikele
Naraminjutikele,
Nikwantalbantitjalo,
Mikwantalbantitjalo

Marankalurknuljele
Mikwantelbantitjalo

Warabitjabitjau,
Waljutulbelo

Ntimankote namarire nalintibe
Ntimaatja ralintibe

Ntimatjibu larintibe
Ntimaatja ralintibe

Ntimankule ralintibe
Ntimanuwo bintjintibe

Nomaarkwe rhkarlanopai
Nomatnjenja lbelanopai.

Red is the down which is covering me;
Red I am as though I was burning in a fire.

Red I am as though I was burning in a fire,
Bright red gleams the ochre with which I have
    rubbed my body.

Red I am as though I was burning in a fire,
Red, too, is the hollow in which I am lying.

Red I am like the heart of a flame of fire,
Red, too, is the hollow in which I am lying.

The red tjurunga is resting upon my head,
Red, too, is the hollow in which I am lying.

Like a whirlwind it is towering to the sky,
Like a pillar of red sand it is towering to the sky.

The tnatantja is towering to the sky,
Like a pillar of red sand it is towering to the sky.

A mass of red pebbles covers the plains,
Little white sand-rills cover the plains.

Lines of red pebbles streak the plains,
Lines of white sand-rills streak the plains.

An underground pathway lies open before me,
Leading straight west, it lies open before me.

A cavernous pathway lies open before me,
Leading straight west, it lies open before me.

He is sucking his beard into his mouth in anger,
Like a dog he follows the trail by scent.

He hurries on swiftly, like a keen dog;
Like a dog he follows the trail by scent.

Irresistible and foaming with rage—
Like a whirlwind he rakes them together.

Out yonder, not far from me, lies Ankota;
The underground hollow is gaping open before me.

A straight track is gaping open before me,
An underground hollow is gaping open before me.

A cavernous pathway is gaping open before me,
An underground hollow is gaping open before me.

Red I am, like the heart of a flame of fire,
Red, too, is the hollow in which I am resting.

When the aborigines eyed my water bag, I gave them a few pints and went on, determined to make Balladonia. At Newman's Rock there was water. It tasted salty and brackish, but was good for tea, a shave and a wash, providing I did not reach Balladonia, where homesteader Jackson would provide me with a kangaroo steak and enough water to take a bath. With the light on the scooter faltering, I decided to make camp, though I was only twenty miles from Balladonia. I gathered loose mallee timber and built a roaring fire. I made tea again, tossed a blanket over my sleeping bag to keep off the heavy dew and I went to sleep. When I awoke, four hours later, I had company —two young truck drivers who had pulled up to get warm.

"We're having trouble," said one of them. "The brakes are going fast. And with eleven tons of spuds, which will bring twice the price in Adelaide, we'll have a lot of potatoes to eat if we don't get cracking."

"We're only making fifteen miles an hour," said the second driver.

"Can you fix it in Balladonia?" I asked, getting ready to leave.

"You can fix nothing there; but you can get petrol and meals. Be seeing you," said the first one, as I scooted off.

My palms were burning. I had clutched the handle bars almost around the clock, and every bump ran painfully through me. My eyes were sore and reddened from the powdery, fine dust. When I reached Balladonia, hours later, I was ready for a rest cure; instead, I got steak and eggs. My khakis were a fine shade of deep brown, and my scooter looked as if it had undergone a camouflage, but with nature's own Eyre Highway colorations.

"Well, I don't know," said Mr. Jackson at the Balladonia Homestead, looking me over when I arrived. His place was a ramshackle affair, put together with corrugated tin and waste timber. But it was a place between the vast wastelands.

One hundred miles south of Balladonia was the Great Australian Bight. The next fueling depot was at Madura, 196 miles off, where homesteader Anderson and his family ran a small, rickety hotel. Normally my scooter could do 250 miles a day on fair roads, but at ten miles an hour I saw myself fumbling and tossing wildly for twenty hours, providing I did not rest.

Two planes passed over. I suddenly remembered that an atomic explosion was due today at Maralinga, the British proving grounds deep in the Nullarbor desert. My mind, somewhat radioactive anyway, suddenly became acute with explosive nuances. Now I had atom bombs to face in this dry, dead earth, with only the saltbush and the sand, and I visualized atomic clouds mushrooming over me. I was sore and afraid, thinking of the Biblical end of the world.

The desert desolation, with the occasional uprooted mallee trees, now took on human forms. The shapes were things on a Daliesque landscape, melting into my burning eyes, joining everything up in mirages. On this plain of nothingness, everything began and ended with roads. The telegraph line was communication to complex civilizations. I heard the lines humming, singing, orating, making speeches to me, their coppery gleam suspended from the towering sun that overshadowed, for a quick moment, the saltbush wasteland.

The drone of the planes was still resounding when I came to a gate that closed off the road. Cattle were moved through these powdery ranges, and the gates controlled their free movements. Thirty-six miles later, I reached a thousand-gallon water tank, or so said my map. It was empty, as usual. The next tank was at Cocklebiddy, a hundred miles farther down the Eyre Highway. What, I wondered as I drank sparingly from my water bag, would I do when my water was gone? Madura was still fifteen scooter hours away, at least. Happily, it was still too cold and my water might last. In the searing heat of an Australian summer, I would really have seen mirages on the Nullarbor. I gulped more water and thanked the spring season for my lack of thirst.

A westbound car, containing three wool shearers, stopped. I was told in juicy Australian slang that I'd never get through on the scooter at a place called Eucla, 150 miles away; that unseasonable rains had bogged everything in sight; that their car had been stuck for three days before the earth hardened enough to give them a surface; that a report had come over the radio that it was raining there again. When I went on, I was convinced of my foolhardiness in attempting this crossing, using some choice American expletives at nature's indifference to me and my little scooter.

I quit for the day or for the night around midnight, at Moonera, which had another empty water tank. I shook like a man with palsy as I looked in. My hands were laced with blisters and I staggered about in drunken fatigue. I made a fire and fell into my sleeping bag as though I never expected to get up again. I was a beaten man, with new and old pains shooting through my body. But I could not fall asleep, for every position had its difficulties. A bath or a swim would relax my hopelessly aching muscles.

I climbed up the water tank again. Apparently, I had grown used to seeing nothing in them, and my eyes had fooled me.

There was water, but it was brackish. I undressed and lowered myself into the polluted water. I kept my hands up, for my broken blisters might get infected. For fifteen minutes I luxuriated in this polluted Luxor, imagining myself in Tokyo, where a young woman attendant had once bathed me. A few minutes later I returned to my campfire and fell into a deep sleep.

Madura was an oasis in the grim desert. The proprietor, Anderson, called it a motel, but to me it was an Arabian palace in the Australian bush. I shaved with hot water and I showered with cold water, for Mr. Anderson had full water catchments, and I ate steak and eggs.

Nearby, a touring dumb couple were busily talking in sign language. Finally, the man come over with a pencil and a piece of paper, and he wrote, "How is the road from here?" I answered, scrawling impatiently, for my fingers were too sore, "It's lousy, and it's not a road. It's nature on the loose." The dumb man showed my answer to his dumb wife, who joined him in a grinning reaction to the dubious pleasures of my description.

Everything was hauled from Perth, or it came via the railway that ran to the north, some hundred miles away. When the catchments were empty and Mr. Anderson had to haul his water, then no one showered or shaved. They were fortunate even to have tea with their very costly meals. Besides, the hardy travelers carried their own food just in case the motel had nothing on hand. Madura lay a mile off the Eyre Highway under the protection of low hills, out of the reach of the winds that came up from the Bight and rolled across the Nullarbor plain in great velocity.

"The atomic explosion has been called off," said Mr. Anderson as I fueled up. "Too much wind for the fall-out, or some such scientific reason has been given over the morning wireless. Lucky for us, eh?"

The aborigines living within the radioactive radius of Maralinga had been moved away weeks ago; but the winds could not be moved away. Millions of dollars had been spent preparing for the big blowup; and all the trappings, from dummy cities to dummy humans, were grouped around Maralinga. I was a safe distance away, at least I hoped so; but knowing how the winds tended to change without warning, I hurried over the sandy gloom of the Eyre Highway, just in case an ill wind, now only blowing up the thick sand, decided to blow the heavens in my direction after the explosion.

Eucla, the next listed stop on the Eyre Highway, was a mere 112 miles away, and there a Mrs. Gurney had contributed thirteen children to the Nullarbor's scarce population. Apparently there was nothing else to do but bear children while waiting for stray humans, wandering madly over the desert, to come and ask for petrol and food.

The two truckers arrived just as I was leaving Eucla, having driven all night in low gear. They looked bedraggled and fatigue beaten, staggering from the transport and almost falling to the ground. Mrs. Gurney, in motherly fashion, was soon preparing a hot breakfast. This time I insisted on hot cakes with my steak and eggs.

"It's not worth the four hundred pounds we're getting for

carrying this load of spuds," said Bluey, the tall driver. At two dollars and twenty-eight cents to the Australian pound, they would make nine hundred dollars for the job. It sounded like gold-rush days, with pay dirt exploding all over the Eyre Highway for men who could push through from the Indian Ocean to the Southern Ocean with the lowly spud.

"Probably see you at Ceduna, where the rains are heavy," I said. "I may want a lift for fifty miles."

"If we still have this bloody truck, chum, you're more than welcome," said Bluey, sitting before his breakfast as if he were in a trance.

"Gentlemen, the tea's getting cold," said Mrs. Gurney.

The next stop, which had neither water tanks nor primitive accommodation, was White Wells. The road passed through mixed scrub country and smelled of the sea, for the Great Australian Bight was not far off. The wind had risen and it was hitting my wind-screen, making it difficult for me to hold the road. Occasionally I wandered into the bush, edging over mallee trees and crawling lace lizards, their fat little bodies squirming to get out of my way, so that twice I took headers to keep from killing them.

I was at the tail end of the Nullarbor Plain, that enchanting emptiness that was, nevertheless, the grazing ground for brumbies, wild pigs and domesticated sheep, eating the spare grass that managed to defeat nature's aridity, growing in dull green shades, as if in protest.

I passed places that had beautiful aborigine names to identify their waste architecture in the Nullarbor; and each had a distinct sound that seemed to come from the bare earth itself. I wondered what places without human beings or villages meant —but I found a peculiar magic in names like Bunabie, Albala, Koonalda, Guinewarra and Yangoonabie.

I reached White Wells late at night; alone again, camping near an empty homestead, frying two pounds of steak and breaking out a bottle of South Australian burgundy. The wine was warming, for the night was bitterly cold and damp. I roasted and froze at alternate times, then moved a few feet away. Ahead of me was a small mallee tree that someone had dragged onto the road to indicate a large, dangerous pothole. Tumbleweeds, or roly-polies, blew over my sleeping bag into the flames. Here was the dead heart of the Australian bush or the never-never, as the aborigines called it. One day, with migration from Europe to Australia building the country, people would come and roads would grow and the Nullarbor would be like parts of California. The frontier would vanish and only nostalgia would remain.

White Wells was 144 miles from Penong, with the worst of the road still ahead of me. Peculiarly, the closer I came to civilized places, the worse the road became. Some hardy people had made a little village 1,180 miles from Perth, and the first village after Norseman. Penong was the beginning of good wheat and pastoral land; but the traffic had increased and the ruts and the rains played havoc with me. I moved through thickly padded mud lanes, constantly forcing the vehicle from the left to the right side of the road. When hopelessly bogged, I got off but kept the motor running, forcing the scooter ahead in low gear. I was General Mud of the Nullarbor, but somewhat in a hole or getting out of one. When I reached higher ground and a few miles of comparatively dry road, I raced at fifty miles an hour to make up time lost in the bogs of the Eyre Highway.

The scrub and the mallee trees were giving way to large trees, to pastoral images that warmed my vision but did not help the scooter very much. This was no longer Western Australia, but South Australia. If I reached Penong, if ever I got there, I would celebrate my forty-ninth birthday, providing I got there before pub closing, which was 6:00 P.M.

A car passed and I waved it on, as if I were a traffic cop worrying about the sudden crowding on the road. But the car stopped, either by desire or because the mud dictated the stop. The driver was a local salesman, a traveler, as the Australians quaintly call their salesmen. He offered me a lift, though he had no solution for my scooter. How would that be transported? I asked with amusement. I wanted to celebrate my birthday in Penong, which was still miles of mud away. I laughed off the hitchhike suggestion and watched the car plow slowly through the mud, wondering how soon the axles would go, and I, some hours later, would have a salesman hitchhiking on my scooter.

The country was changing, with human touches now evident. Odd signs appeared, one saying, "Normanville, 850 miles away. Beer, Bed and Girls."

No doubt, it was a come-on to make the weary pioneer forget what he was going through—and I was going through mud that never, without forcing the language, became sensual. I saw desolation give way to isolation, vegetation and possible radiation. Had the bomb gone off at Maralinga, the field of thunder? I'd know about that soon enough; but now I was mucking through the lower depths of the Eyre Highway, thinking of the postponed private birthday party I was going to give myself at Penong—or would it be at Ceduna?

I got through, but Penong was locked up solid. The hotel was shut and the garage as well. The place was dark and I was terribly tired, the pain in my palms and body stabbing wildly. But I was very happy, feeling a secret joy I could not fathom rising and welling at my emotions. I had tried something at forty-nine; I had challenged a road around the world. But I fumed, for I wanted to celebrate myself in Whitmanesque attitudes.

Fifty miles before Kimba it had rained for days and the mire and the mud was knee-deep. I would have to quit there, for it was impassable, and my anger seemed justified. I decided to go on, to head for Ceduna, forty-four miles away. Ceduna was a real town, complete with a movie house. It also had a proper hotel, I had been told. I desperately wanted a bed and not a campfire along the road. I was exhausted physically and my eyes were getting bad. I remembered a doctor in New York telling me that I had a dust and wind allergy. I should see a doctor, I thought, using the last of my water to wash my eyes and face.

I saw a campfire, with a group of male aborigines about it, their *lubras* (women) sitting cautiously in the background. I stopped, uninvited, making a rather grotesque appearance alongside their fire, my scooter humped against a mallee scrub.

I was the strange outsider, the man from nowhere looking for a birthday to celebrate. But my water bag was empty, my language not at all Australian; and I looked all the more curious as I flopped down and had one of the women give me a *cuppa*.

Embarrassed, I broke out the leftovers of my cognac and passed it around, watching the reaction of the three men and their three women. After much hesitation, one man finally took it, smelled it, then said, "It's not beer, chum." He might well have added, in similar mission-learned English, "It's also slightly illegal, chum; for we are aborigines and only the white man is allowed to get tight."

But no one got tight, for we had the cognac with tea, followed by a brief corroboree when I told them that tonight was my birthday. They broke into a dance and I tried to follow the complex ritual steps. When they finished, I tried to entertain by doing a *kazatzka*. I made a few wild leaps in the air, then found it impossible to continue, for my muscles had tightened up, ending my Russian dancing. But I'd had a birthday in the bush; and with the cognac gone and the tea drunk, I went on, bidding the aborigines many happy returns from nature and man.

The road was now hard in part, though the potholes were deeper; but at least I did not bog as I whirled from one hole to another, looking for better surfaces. My vision, with the scooter light blinking idiotically on and off, as if it had delerium tremens, was blurring, enabling me to see only a few feet ahead. Then it happened . . .

A kangaroo bounded from the bush, leaped across my path into a pothole, where it stood transfixed by the light. I hit it as I bounded up, the sharp edge of my front baggage-carrier cutting into the kangaroo's neck. I hit the ground, entangled in the kangaroo's slashing feet, then jumped away to keep from being struck by a death-maddened foot.

I was bleeding badly, with the old Ceylonese wounds opening up. The kangaroo was twitching and pawing helplessly, bleeding from its neck. Soon it lay still, its blood running into the pothole.

I was confused and conscience-ridden for I had killed the

kangaroo now lying laced about my scooter. I pulled it to the side of the track, looked my scooter over for damages, started the motor, then found that I had tears welling up. I was indeed the sentimental traveler, not hardened enough to face the death of an animal that might well have killed me instead.

This was the middle section of my bizarre birthday, and I was attending a private wake for a kangaroo. In any event, I could not go on; for my knees buckled with pain when I tried to ride, the iodine I had put on burning me more than helping my situation. I got out my sleeping bag, made a fire from the mallee scrub, but could not fall asleep.

I saw kangaroos jumping throughout the night, bounding over me, through the fire, their bodies bloody in their streaking flight within my fancies. They took on human shapes, looking like old friends; one even looked like Emily, a girl I once knew in Melbourne, with whom I had committed an ancient indiscretion in the year of our Lord, 1947, and I vomited, endlessly accused by my conscience, revolted at this strange kind of death I had caused. I had been through the World War II, but killing animals was still new to me. I slept.

I awoke, shivering with anxiety and cold. The fire was ashes. I wanted tea but my water was gone. I had used up the last at Penong to wash my eyes. I shivered inwardly, as though I would never get warm again.

A mist hung over the early morning on the edges of Nullarbor. I put on three more shirts and found my old gloves, the hard old leather chafing the tender blisters on my palms.

I took my camera and walked to where the dead kangaroo lay. The crows had begun their feast, but they flew off hurriedly, whining their objections so that I became conscious that now, too, I was completing a design against nature by depriving them of their natural feast. Sadly, I photographed the dead kangaroo.

A mile from Ceduna I saw the first aspects of modern Australian civilization—the racecourse, then the hospital. I drove to the hospital, entering with a state of nerves, my knees not for Russian dancing. A nurse put on new bandages, then informed me that I had almost burned myself with the iodine. My explanations seemed to amuse her, including my ASPCA spirituality. Apparently another dead kangaroo was a statistic, often too fatal for others on the Eyre Highway.

"You were very fortunate," said the nurse as I was leaving. "You're alive, with some very beauto bruises. How fortunate."

But I wanted more fortunate things, like a real bed and a bath, and I got both in due time at the local pub, where an agreeable Mr. Watson, over a late morning drink, began regaling my tired mind and nerves with fishing stories. Ceduna, like Port Lincoln, was a fisherman's paradise, with the best whiting in the world romping about in the offshore Southern Ocean. Did I know, too, that the novelist Zane Gray had written some of his best stories about man and fish at Port Lincoln? I did not; nor could I bow out, not after Mr. Watson learned that I had celebrated my birthday in the bush of the Nullarbor.

He broke out a pinch bottle; but I begged for steak and eggs, promising to consider the merits of the bottle later in the afternoon after I had slept and washed away some of the desert dust.

"Much better if you went fishing immediately," insisted Mr. Watson, doing quite a job with the bottle. "As a matter of fact, you are joining me in a boat almost immediately. The cold, bracing Bight wind will fix you up."

I ran to my room and Mr. Watson went fishing alone. I showered my unwounded parts for thirty minutes and rubbed myself down with whatever vigor I still had left. I plowed into a soft bed—awakening when I heard familiar voices. It was Mr. Watson leading the two Australian truck drivers to my door; but it was also ten hours later.

"Now we'll celebrate," said Mr. Watson. "These two gentlemen have also managed to come through in one piece."

"You ought to see the truck," said Bluey, the tall driver. "It's almost had it."

"Or us," said the other, swinging a glass in front of me. "Here's to a bloody fine Yank. Happy birthday!"

"To many more forty-niners," gagged Mr. Watson, who immediately began praising a local fisherman who held the world's record for size and weight of sharks caught. "He's hooked the three greatest, biggerest in the world. Why, Sir William Penney, the bloody Pommy atomic scientist, was here the other day and he even caught a large one. This place is foul with big fish. So here's to you, Yank."

In Australian sentiment this was called mateship; and it had all the echoes of the American frontier transplanted in place and language. Everything they said glowed with goodwill and wit. I was a stranger they had met on the wastes of the Nullarbor. But it was my belated birthday.

We relaxed with more fish stories. In the comfortable brown lounge of the pub, Mr. Watson brought out his trophies, which were not all pinch bottles, but gold and silver medals for real fish that he had caught in competition.

I was not a fisherman. I merely ate fish. So at dinner we had a whiting course; and Mr. Watson was even more loquacious. It was excellent whiting and excellent beef and terrible coffee and beautiful South Australian wines. A pilot from Ceduna's Flying Doctor Service offered me a ride, too. I had a scooter, I insisted.

"Leave with us in the morning," said Bluey. "Cop a lift for the next 200 miles. The weather's going to be foul, and the road from Wudinna to Kimba looks like its been invaded from Mars—just one long bog for sixty-three miles. Come on, Yank."

A few cars had come through, looking as though they'd gone plowing in a rice paddy. But it was 142 miles to Wudinna, miles that I scooted, with a promise to meet the truck drivers at the Wudinna garage within twenty-four hours; that I would join them for the sixty-three miles of impassable terrain from Wudinna to Kimba.

Ten miles from Ceduna I passed a typical fence put up to keep the wild dogs, the dingos, from overrunning the sheep land and feeding off the sheep. There were rabbit fences too, though the Australian rabbit was no longer a pest, for myxomatosis had killed off millions of rabbits a few years back.

With a shudder, I saw a dead brumbie, the wild horse of

Australia that usually ended up in Singapore, when alive, for work in the fields; and I saw three dead kangaroos within a few miles of each other, as well as two broken-down trucks loaded with potatoes. Mayhem and accidents on the Eyre Highway were universal, apparently.

Many migrants from Italy had come to this region, quickly acclimating and losing their strangeness in looks and language. At the Wirrulla garage, fifty-eight miles past Ceduna, I heard a young Italian mechanic speaking Australian slang, though he had only been in Australia for two years. I kept listening to his admixture of Italian and Australian, to the overweighted cockney accent so many Australians delight in using. When I heard the Italian mechanic say "Pronto, cobber," in the most exacting Australian accent, I knew that he had become a citizen, and that he was very proud of it.

The grazing land now stretched to all the blue directions of the horizon. At the garage in Wirrulla, I had heard a newscaster say that the atomic explosion set for the late afternoon at Maralinga was once again postponed; the winds were still wrong. Yet right or wrong, whenever a car passed, which was about every three hours, a cloud of brown dust would whirl up, choke me and make its own mushrooms in the air. Had the traffic been heavier, it would have been impossible to see at all, or I would have choked from Mother Earth's eerie highway. When dry, the road was a whirling brown cloud; when wet, it was a meandering bog. The choice, though there really wasn't one, was the dry, acid-tasting, windblown road.

Nearby was the turnoff for Streaky Bay, a fishing center on the Southern Ocean, much easier to reach by boat or seaplane than by road. I was now in cattle country, and houses showed up with more frequency, one every fifteen miles. The range riders rode planes over this vast grazing land, spotted the huge herds, then radioed back, and men with motorcycles, dogs and packs cruised over the open country, rounding up the cattle for the drive to the railroad, hundreds of miles to the north.

This forbidding land, especially around Esperance, running along the Archipelago of the Recherche, would soon find settlers; for Americans had bought up over two million acres to grow rice for export to the East. It was an industrial project, exciting others who saw a future there if one had money to work with. American investments had grown to seven hundred million dollars in Australia, now considered economically safer for industrial expansion and speculation than any other country. But my own speculation was whether I would get through, after Wudinna, unless I met the two truckers to cart me over the bogged road, which their wheels could take and mine could not on the terra non-firma of Terra Australis.

Twenty-four miles before Wudinna, where we were to meet, I stopped for gas. The woman attendant, somewhat frayed by time and work, appeared anxious to make conversation. Her husband ignored me and let her turn the gas crank, during which time she kept repeating, "I don't know why I stay on in this place. It is a sinful, terrible, God-forsaken community of twenty people. Why, the white men around here live with the aborigine women; and they are not married to them. It's terrible, sir. It's sinful and terrible, living without benefit of clergy with those gin women. I don't know why I stay on in this sinful place." She paused, sighed, gave the crank a final turn, then said, clear out of the blue, "Why, even Bill and I are not married—it's terrible."

Bill, her unmarried man, merely grinned, feeling sufficiently sanctified by that effort. Since I could not very well answer or even smile back, I paid up hurriedly and escaped my amoral company.

I reached Wudinna after dark. At the garage I was told that the truck drivers had phoned through; that they had broken down but would make it by morning; that I was to sleep over and meet them for breakfast.

The truck drivers arrived early in the morning and parked in front of the hotel, their truck taking up most of the street. When I arose I saw two very weary Aussies asleep on top of the truck, their feet protruding over the truck's sides, their bodies and faces covered by the heavy tarpaulins.

The truck itself looked like a scarecrow on wheels, with baggage strewn over its length and width. The tires were shreds; the fenders were caving in; the radiator had a deep slit across it; and the springs sagged, giving the body a curvaceous design. Such repairs as they made were always temporary; for it was not worth putting money into this ancient affair on wheels. When they heard my greeting, they poked their faces from under the tarpaulins, grinned, and jumped off the truck. A week ago they had left Perth on their first trip to Adelaide. Now, grimy and worn with fatigue, they could only grin. But the grins meant nine hundred dollars and a downpayment on a new truck.

We left with grinding gears and brakes that seldom braked. I sat on top, along with two young English hitchhikers they had picked up. My scooter was braced on the tarpaulins, jammed between edging sacks of potatoes—then strapped down with ropes and cables. Now the ride to Kimba on General Mud started.

General Australian Mud began a mile out of Wudinna, where the rains had been much heavier. The road looked like a battlefield plotted with deep furrows, which kept trapping the huge wheels of the ten-wheeled transport. The soft, slithery Eyre Highway was a morass of all the elements juicily gathered together. We passed two local trucks during the first five miles, axle deep in the oozing earth, where they would stay until a tractor came to haul them back to Wudinna.

It was painfully slow going for the first three hours. I tried sleeping, but with the sudden stops and lurches, I almost went over the top and into the muck. The two English hitchhikers were ex-sailors who had made their way from England via the rule of the thumb; and having used up what little cash they had during their seven months' journey, they were planning to take farm jobs at Kimba. One of them had an almost pure Oxford accent, which he delighted in using on people, perversely enjoying their chagrin, especially since he looked like a bum. A few weeks before, in Perth, he had knocked on a door for a handout. The lady turned him away, saying, with linguistic irritation, "I hite anybody what cain't talk gud Inglish.'

"Join the Australian Labor Party," I said. "You have a talent for comedy."

But so did the transport have talent; for it was a feat of

engineering, via Scotch tape, which held it together. We held together without Scotch tape, hanging onto the ropes that laced down the huge tarpaulins covering the potatoes. When sleepy and cold, we burrowed beneath the tarpaulins. At one point the Englishman with the Oxford accent rolled deep down into the tarpaulins and we had to scoop him out of his little canvas igloo.

At Balumbah Siding, with only seven miles to Kimba, we were swamped as we splashed in the ever-ever-muddy ravages of the Eyre Highway. Finally, we made it to a side road that ran to a place called Waddikee. Pausing, we held a conference with some bottles of beer and sandwiches. Kimba was five miles away; after that the road improved, with the Eyre Highway ending just before Port Augusta, where a good bitumen road ran the last 250 miles into churchly Adelaide.

It was a victory over mud, though we had mud in our eyes and over our clothes. We toasted our beers to Terra Mud and started off again. Bluey, the tall driver, gave the wheel to his shorter colleague, and the hitchhikers and I scampered back to the perch on the tarpaulins.

We rolled or lolled into Kimba fifty minutes later at six miles an hour. The road changed with magical suddenness. My little lift had come to an end. I could manage, somehow, on what was left of the Eyre Highway.

A railroad spur ran to Kimba, and bales of wool were collected there for shipment to Adelaide. A dozen huge bales were piled in a field near the railway tracks. A fence ran around the railroad property, where the Eyre Highway took an abrupt right turn. A faded sign warned drivers of the sudden turn, but you would have to know about it to read it; and tall grass grew between the railroad fence and the sharp turn in the road.

The driver, happy to see a hard road again, gave the transport the gas on the stretch before the turn, for neither the sign nor the turn showed up clearly. Before he could check himself he was at the turn, but he was not turning. The transport broke through the fence, ran fifty yards into the railroad property and smashed into the bales of wool.

Down we came from our perch in one fell drop. The transport buckled, making a ghastly mechanical discordance. The cab was stove in and the radiator split in two. The right front wheel was hopelessly bent; and a log, from some unseen nowhere, had come up from the ground and slashed into the front springs.

"This is it," said Bluey. "We've finally had it. Oh, well, we're insured, so let's take stock of the damage."

The local garage owner wobbled over. After some laughter, for no one was hurt, he said, "You can use one of my trucks to haul the spuds to Adelaide. I'll only charge you eighty quid providing you do the re-sorting and repacking."

"It's a go, mate," said Bluey. "Let's all have a go at it."

We worked through the day shifting the 180-pound bags of potatoes from the broken-down transport to the second transport, which was parked outside the railroad property.

A cop took down all the facts; an insurance agent, more facts, during which time we labored in the railroad yard with the spuds. Never had eleven tons of potatoes weighed at least eleven tons, we agreed.

With the potatoes shifted, I scooted on to Iron Knob, reputed to have the largest ore deposits in the Southern Hemisphere. The deposits were hills of iron, literally; for all the hills around Iron Knob had been scooped up and trundled off, the remains looking grotesque in the sun's afterglow.

I was now approaching the industrial heart of Australia, with the Eyre Highway's bogs, tracks and sand drifts giving way to bitumen road; the dusty, impassable land magically turning to pastoral scenery. A sign, placed dramatically, with typical Australian understatement, merely announced at a fork in the road: *Eyre Highway to Perth 1500 miles; Adelaide, 250 miles.* I went toward Adelaide, looking back before I took the final turn at the worst, most dangerous, longest, most Godforsaken, death-defying road in the world, U.S.A. 1 and July Fourth notwithstanding.

I was on a royal road, going forty and fifty miles an hour instead of ten, with the gales from Spencer's Gulf blasting into my wind-screen. But it hardly mattered about wind and rains, for I could move again, and my natural exultancy got me to Adelaide in seven hours. It had been eight grueling days and nights for man, scooter and beast, with the warm Australian mateship of truckers, pub owners and aborigines making nature's own highway a companion of spring. Had it been summer, it would have been worse than Iran; a form of self-induced mayhem, sans Persian brigands and cutthroats. But I had come through in both places, and so I celebrated that which every frontier gives up to every man, his indestructible spirit to accept and complete a challenge.

Like the rest of Australia, Adelaide had grown by a full head, adding the complementary girth to its waistline. The "village," as it was quaintly called by visitors from Sydney and Melbourne, had sprawled directionless, almost doubling its population within ten years. The lovely Torrens river, into which Max Harris, once the *enfant terrible* of Australian poetry and letters, had been thrown, looked even more soothing as Max and I walked along its gentle banks and watched oarsmen pull at their single sculls. It was like an old print except for the missing foxes, red coats and hounds.

Max Harris had grown too; for no longer was he the editor on whom the famous Ern Malley poems had been hoaxed during the war; he was, like Adelaide, burstingly corpulent, with the good life of poetry giving way to the better life of a successful bookshop owner. He was so successful he sneered at poets or prose writers, and even painters, who did not have bookshops to grace their late thirties as a way of life. He made trips to England and Europe every two years, and could have stood for Parliament, physically, had it not been for the Torrens river and the long memory of the students who had heaved him into its waters for alleged radicalism in economics and poetry.

Harris had arranged to get me a room on Childers Street in North Adelaide at a house that starred, in various libidinous roles, gentlemen and lady amateur actors, who were more professional during the day while working as clerks, carpenters, photographers, porters, or salesmen in other congenial arts and crafts. At night, when not rehearsing or playing to Australia's love for things amateurish, the assorted roomers roomed with each other, and their quaintness as loose lovers was quite exacting when the tumult and the shouting began, which usually went like this, after midnight: "David, who's the sheilla you're poking to sleep?" Or, "Where's that bloody wife of mine?" a male voice would utter. "Has she made the rounds of all the bloody rooms, including that bloody Pommy's and that Yank's?" The echoes following would have shamed Rabelais, who was not an amateur in matters of dalliance—but this was amateur night in North Adelaide, in a play that probably is still running, with visitors being made quite "to home, with a cuppa tea."

Getting off the scooter to lecture to Adelaide's Fellowship of Australian Writers on the "Pleasures and Horrors of Contemporary American Literature" was a change for the better, physically and spiritually, despite the peculiar subject matter. Prior to lecturing I was interviewed by the local press and radio, which helped to bring an audience of one hundred writers and nonwriters, with the typical spinster type, horn rimmed behind shell glasses, taking hurried notes of my easy course to pleasure and horror, my study of gutter publishing riding America's literary wake and nightmare.

I must have sounded like a moral crusader, which is an odd role for me, as I delineated the horrors of the infantile, murderous, breast-bulging stew the publishers offered up in drugstores and without tranquilizing pills. With Australia following America's commercial personality, importing the garish gadgetry from our assembly lines, I was indeed filling a strange role for Judgment Day in matters of the mind and the body.

"To Horse!" or better, "To Scooter, and Ride!" or "Take to the road and not to the bars, get healthy via a scooter trek around the world," seemed to be my antidote for America's contribution to the publishers' garbage can of values. Three hours and a hundred questions later, I saw myself heading a caravan of poets on scooters, the Minnesingers heralding the flight back to reality—out of the garbage and onto the road back to moral meaning. The American outhouses of the publishing industry had had one grand swill of a time, and we called it an evening, after midnight.

It was a month of lectures, with talks to the North Adelaide Junior Chamber of Commerce on life in Turkey and Iran; over the Australian Broadcasting Company networks on American poetry and American politics; and to the university students, who were very critical in their quiet way as I read from my travel journal. I ended my dozen assorted lectures by talking to the Adelaide Scooter Club on how you can go around the world for a thousand dollars, which is a lot of money if you have none, but nothing to see a world on. It turned out to be an exciting evening of machinery and boys; for we were down to gears, gasoline costs, spark plugs, sleeping bags, medicines needed, guns not needed, maps that were correct, languages that you did or did not need, and the question of companionship. "Don't go with a blonde girl," I said, remembering Peter Winant's murder in Afghanistan . . .

I began to exploit my trip. My scooter was in the window of Bruce Smalls Ltd., who ran the Vespa agency in all of Australia. Alongside of my scooter was a garland of gallon-cans of Mobilgas gas, along with a dozen photos of me in different parts of the world. For this little display I received from Bruce Small Ltd. and Socony Vacuum my fare back to the United States, very happy to give all to the pictorial exploitation of their excellent products.

I had come upon this idea one day in the Persian wastes, which was a good place to think of money and the good life. I even had myself photographed drinking gin, thinking that my agent in New York would manage to sell this to a liquor company. I sat on a sand dune wondering whether I ought to add a black patch over my right eye, though I had my havelock fluttering in the Persian winds. I was the Compleate Poete, though hardly Elizabethan, thinking up commercial ways of exploiting the scooter and me, when next I reached civilized places; and Adelaide was very civilized, especially Bruce Smalls Ltd. and Socony Vacuum. I felt renewed on the last frontier of Western man's gallant ways, loving Adelaide to the last drop of beer and gin, which I never ever drink at all.

Yet Adelaide, so churchly and stately, rushing toward modern destiny, was still in its swaddling clothes culturally. It had some of the minor visual images that Paris gives off in great rushes of excitement, for many espresso cafés had prettied up some of the barren streets, where one-story buildings populated an architectural death on the city. In the cafés young men and girls were aping juvenile America in clothes to fit their sartorial rebellion against the status quo. Teen-agers of all ages had formed battalions of exhibitionists, calling themselves, in Australian patois, Widgies and Bodgies. The boy rebels were the Bodgies, with the ladies wedging in. They made riots about rock 'n' roll films; they smoked marijuana and took opium; they had illegitimate children, which is common enough in Anglo-Saxon countries to make them legitimate enough, as they acted out their teen-age rebellion against abstractions that offered them nothing but sensations. It was an empty display, idealess but energetic, with a complete lack of mental challenges or political and social meaning. They had accepted young America's dead-end rebellion and the dead-end claims, with the assurance of security, which their fathers aimed at solely. And since securing security seldom brings that alone, for it is a killing process, as we have discovered in the United States, they were aimless, mindless, secureless and challengeless, as they energetically raced through their exhibitions of frantic, infantile fancies.

The more arty ones, however, embraced the perverted sexuality that went along with this abstract rebellion; for lesbianism and homosexuality were rampant, with career girls leaning toward perversity as if it were the acceptable bed to success. And churchly, dignified Adelaide hardly knew that behind its stately facade an unholy brew was stewing and rampaging, lewd, nude, libidinous and perverted.

One Sunday, Professor Norman Jeffares, who had come from England five years ago to teach English literature at the University of Adelaide, and I motored to Anlaby to poet Geoffrey Dutton's ancestral estates. It was a tour through South Australia's famed vineyards, with Seppetsville's lush grapevines flagging us on to stately nineteenth-century Anlaby, where Geoffrey's mother reigned like Queen Victoria, if with a more terse tongue to add to her many charms and graciousness.

Geoffrey was a fine poet, whom I had met during the war in Sydney when both of us had books published by the firm of Reed and Harris, also the publishers of the realist, surrealist, naturalist magazine, *Angry Penguins*. The magazine had originally printed the hoaxed poems of Ern Malley, the phony poems that James McAuley and his friend Stewart had concocted in the mosquito-ridden swamps of New Guinea during the war. But Anlaby, somewhat removed from New Guinea swamps, soon had its Georgian atmosphere ringing with mad laughter about the famous hoax. I, too, had been taken in by the hoax; for I had edited, with beautiful Liz Lambert, a Sydney poet, an Australian anthology that was published by a well-known magazine in the United States. Liz and I took Max Harris' word that all was practically above board when he gave me the poems to read; that the poems were authentic; that the alleged author, Ern Malley, had died of Graves disease; that Ern Malley—or so said the mysterious letter that introduced him and his poems—had died when his body exploded due to a lack of enough iodine in his system.

Max Harris was just the right man to go along with these concoctions; for Harris, thinking romantically, began to proclaim in print that Australia had found a dead, Down Under version of Rimbaud. And with Harris' imagination reaching beyond the normal Harrisian stratosphere of the exploding imagination, he was soon doing a fantastic public relations job in promoting the unknown, exploding poet, Ern Malley, created out of the whole cloth by two Australian soldiers, who were themselves good poets but who disliked modern jabberwocky, which *Angry Penguins* signified to them because it printed Dylan Thomas, Henry Treece, Herbert Read and other alleged obscurantists.

Nonexistent Ern Malley croaked on metaphysically, with the voices of Stewart and McAuley hoaxing on behind, in his "Sybilline":

> That rabbit's foot I carried in my left pocket
> Has worn a haemorrhage in the lining
> The bunch of keys I carry with it

Jingles like fate in my omophagic ear
And when I stepped clear of the solid basalt
The introverted obelisk of night
I seized upon this Traumdeutung as a sword
To hew a passage to my love.
And now out of life, permanent revenant
I assert: the caterpillar feet
Of these predictions lead nowhere,
It is necessary to understand
That a poet may not exist, that his writings
Are the incomplete circle and straight drop
Of a question mark
And yet I know that I shall be raised up
On the vertical banners of praise.

The rabbit's foot of fur and claw
Taps on the drain-pipe. In the alley
The children throw a ball against
Their future walls. The evening
Settles down like a brooding bird
Over streets that divide our life like a trauma
Would it be strange now to meet
The figure that strode hell swinging
His head by the hair
On Princess Street?

We laughed, we drank, with stately, gracious Mrs. Dutton insisting that we have a picnic lunch in the open; and Geoffrey, knowing his Australian rabbits, thought a few would be fine for grilling in the fields. With dogs, cats and humans off to the chase, soon Geoffrey was rampaging over Anlaby's lovely acres scaring up hidden rabbits, who, for some strange reason, refused to stand still or run slow enough for Geoffrey to shoot them in time for lunch. We had a fine lunch, nevertheless, sans the agile *lapin,* who leaped agilely behind gum trees while Dutton peppered the countryside with his hunting rifle.

Anlaby had been settled in the 1830's, when Geoffrey Dutton's forebears came from England with 3,000 sheep, thus introducing the sheep and the wool industry to South Australia. One hundred and thirty years later, the original 100,000 acres had become a mere 3,000, but capable of keeping 10,000 sheep up to their chins in grass; for it was rich land, blooming and blossoming with whatever makes sheep happy roaming Anlaby's hills and dales, baa-ing over the present but probably only too aware that their own ancestors ate there before someone slit their throats.

    Dutton's father, before settling down to raising sheep in 1906, made the first motor trip around Australia, calling in at Alice Springs and Darwin in his vintage car, which he called "Angelina." The vintage photos he took show every variation of the Australian bush, with its primitive inhabitants; for many a boomerang went Mr. Dutton's way as he traversed the Never-Never of the outback. Geoffrey gave me a copy of the picture book his father published, scrawling across its green pages, "To Harry, another amateur of the overland journey." Geof himself was an accomplished overlander, and had written several travel books, having a few years back gone over part of my route; and, not oddly, we had slept in the same room at Olympia, getting the same anthropological lecture from the historian, cum landlord, of the Biltis Hotel in Olympia. And since maps, even offbeat maps, take poets to the same mental as well as geographical places, he had also visited the tomb of Omar Khayyam in Nishapur, but without the exotic benefits attending the "loss" of my sleeping bag.

    Professor Jeffares, who had written a study of the Irish poet William Butler Yeats, which is an accomplishment for a lively Scotsman, had in his student days at Trinity College drawn a mellowed rendition of the buildings of Trinity and its scholarly environs, making the artist, the critic and the man one with himself, something which comes easier to the British breed of men than it does with our own Yankee talents. Professor Jeffares, youngish, donnish, vibrant, with his speech and wit coming at you like a machine gun in a hurry, literally took over the picnic in the preparatory stage, when hunting firewood was the order of the minute; we hunted and he talked, his crisp voice salvoing with conceits, scandals and gossip; all the stuff that makes a literary picnic unforgettable as a tour de force in jovial malice. He had arranged for me to give a talk to the students during their lunch hour, saying, "Shock them, Harry. Read that poem, 'The Return,' from your new book. They need some American realism jabbed into their beer."

When I talked to the students a few afternoons later, I was the one who needed a jab, and right from the start. I was on parade, as it were, a poet who had almost scootered around the world; but they looked on, indifferent to my safari on two wheels, waiting to be shocked. I did, but only with the word "bastard," which is in the poem, for I was hardly aware that I was facing genteel young women and men. I remember smiling impishly, if a bit smugly, as I saw the result of my unconscious seven-letter word that is so richly Anglo-Saxon, and which you can say, with a bastardized smile, at Australian cocktail parties or over tea and crumpets. If some of the ladies present object, however, you get told off by a gentleman who calls her, behind your back, a "sheilla," meaning a would-be-strumpet going moral on him.

A few days later I was invited to become an honorary member of the Adelaide Sports Club, word having gone around that I had, in my teens, chased after Paavo Nurmi at Madison Square Garden, though I never caught up with Nurmi except when he lapped me in a three-mile race. This made me an ex-athlete indeed, as well as an ex-scooterer; for I had grounded myself and gone back to lesser athletic carnivals, being in the company of poets, critics, journalists, and Max Harris, who combined all things in his nonsacred stances.

It was a month of rediscovering Australian values; some of them measured in pots of ale, others in the strange restraints of the intellectual elite of Australia. When the ex-hoaxer, Professor James McAuley of Sydney, arrived for a series of lectures on contemporary Australian poets, I was to meet him for the first time, twelve years after the Ern Malley poems had been published by Max Harris, who still refused to meet him on his own grounds at the University of Adelaide. I was much less squeamish despite my own contribution at the receiving end of the hoax, the anthology I had included them in; and I found McAuley a sincere, intelligent, liberal-minded scholar, who took time out from his professorial duties to edit the magazine *Quadrant,* published by the Australian Committee for Cultural Freedom, the organization that had sponsored James T. Farrell's travels through the land of Studs Lonigan, Down Under version.

The foreigners in Australia were no longer called Refujews, or Bloody Pommies—a name which has since been reserved for the migrating Englishman. They were being welcomed en masse to populate and industrialize, to come bearing their cultural gifts, to enliven the land, of which D. H. Lawrence had once said in an essay on Australia, "In Australia the birds don't sing, the flowers don't smell, and the women have no virtue." But the birds did sing if you were listening for singers; and the flowers did smell, when you left the hothouse gardens and went into the bush; and as for the last of this engaging trinity —virtue is, after all, a transitory thing dependent on the over-all morality. Consider the American marriage and the attendant divorce rates, with every would-be-divorced husband and wife able to prove the infidelity of the opposite number. If this is American virtue, statistically, then Australia is indeed a most virtuous continent.

On Childers Street, a play was in rehearsal. The sets were for O'Casey's *Playboy of the Western World,* but at the last minute the play was changed; an original play, written by one of the actors, who wrote it between lesser moments of bedroom high jinks, was substituted. It had an air of Tennessee Williams, with Williams' pervasive homosexual theme reduced to vagueness, which is not Williams' style of writing. The playwright was even more vague in his characterizations, confusing his men and his women as if they were one and the same sex. In one scene a girl came out talking like a man; in another she emerged as herself or whatever she was by that time; and in a third she was a young boy in Greek clothing, flaunting her old fangs and her new femininity. All together it was an evening that reduced everybody to vagueness; and as the audience left, one was hardly surprised at hearing old ladies addressed as "Mister" and old men as "Madam, your libido is showing brightly." The Little

Theater movement was indeed little, if hardly moving, except toward the exit, which I could not reach fast enough. But this was not the real art of Childers Street's assorted characters; the gay children of sensations, playing at theater between their daisy chains.

One night, after dinner at Max Harris', I learned how a literary kangaroo court, Australian style, can kick over the traces of joviality. With Harris leading a puckish pack of writers and musicians, I was set upon to explain away Secretary Dulles and his State Department; President Eisenhower; Suez; American failures in tennis; our lack of rigid standards in education; our proficiency in jazz and our bad taste in sun glasses; the "murder" of the Australian race horse Phar Lap twenty years before at an American race track; why Henry Miller was being mistreated by the American critics; what had happened to Auden since he became an American; why baseball was so dull and why Americans worshipped its dullness; and where American art was. Harris managed, however, to recognize American novels as terribly important, despite the garbage-can school of prose in the paperbacks; and when he let up, hardly waiting for my answers, his musical friends took up the kangaroo court method, denouncing America as a Dead-end Unto Itself, having given the world nothing but gadgets and tide-over economics instead of political leadership.

It was a strange dichotomy and melange of accusations, and I expected at least the Negro Question, more pertinent and apt, to get its rightful share of the evening's quaint methods; but then, I might have countered with Australia's cruelty to its own aborigines, who were only 70,000 in number (and not 16,000,000 Negroes, or almost one-tenth of the American population) and scarcely a minority problem. But the chase led along burlesquing paths, taking in the contribution of the United States to the strip tease as an aid to sexual maturity; to Elvis Presley as a phenomenon uniquely American, since no other country had developed this hybrid, torrid, tonsil art as well as our shouting, helling vocalists. "And what about the exodus of Charlie Chaplin, who had run away from the land of the free because he was brave?" asked Harris. Chaplin was the true artist, rising above his real public—and after another Scotch and soda, the kangaroo court called it a night.

But Max had developed into a fine poet, using "bush themes" in sophisticated, ennobling attitudes. His earlier obscurantism, which had made him the goat in the Ern Malley box of tricks, had been discarded and a fat cat of a poet was purring over liver and honey in the successful bookshop he had helped to create. I remembered what one critic had said, verbally: "Max has the makings of greatness, even if he only became a great scoundrel in literature," a statement no doubt revised many times, and affectionately. The poet is often a dull man in his own country without his poetry, and seldom if ever a scoundrel, great or small; though he has been accused of siphoning away his would-be-criminal intensity through his poetry.

Having had many dinner invitations, I paid them back in my own fashion. I cooked on two separate nights for two sets of diners at two separate houses, with the Childers Street mansion of diversified arts literally turning itself into a joss house, with several nymphlike worshippers dancing their concepts of the Persian houri, nakedly at home, behind the walls of the garden. It was a successful dinner, all around.

The second evening of oriental cookery was much more sedate despite Harris' raillery and guying; but with the presence of two physicists, a professor of mathematics, and only one nymphomaniac, the evening became too scientific, closing in on itself when the mathematician unrolled a plan to settle the Dead Heart of Australia, with water. "Australia can do what the United States did in the West and the Southwest; it can bring in water. A dozen Great Snowy Dam schemes to develop the water of Australia is a hell of a lot more important than the Australia White Policy to keep Australia pure as the driven snowy."

One of the physicists gave facts and figures: so much rainfall; so many hidden sources under the ground; so many rivers, droughts, bushfires. And he summed it up, "Australia will develop when Australians develop and mature in their treatment of nature. We've abused the land, harnessing it to quick profits from our sheep and wool. Now look at the land!"

Harris summed up Harris, saying, "Look at America—the dust bowl of progress."

I summed it up as the host in a strange house: "The sukiyaki, gentlemen, will not wait for Australia and Australians to develop—let's eat."

We ate amid the bombs and the bombast of physicists dueling over data, a mathematician quoting the niceties of facts, Harris punning, "Roskolenko should be the expert on land problems, what with his travails to get to us," and the silent nymphomaniac, who studied her prospects amid so many scientific minds after science, going to bed.

With the Olympic Games approaching, as well as my departure for Melbourne, Harris took me to see a skull-cracking ballet called Australia Rules Football, a game that is an astonishing combination of American football, soccer, rugger and ballet. It never stops dead and humps like American football, as if each player were on a bender or selling steel to his opposite number, but dashes with startling energy and color. It is not always brutal and neck-breaking but is often politely aesthetic in its gambols on the green playing fields of Adelaide.

A few days later I was scootering to Melbourne, my month of literary living having softened my view of the often harsh Australian landscape. It was a two-day run through gum-tree country, past floods along the everflooding Murray river, where New Australians got their first taste of nature being natural as their new homes were washed away. I saw the ugly ruins left by too much water; and I remembered that Australia suffers oddly from flood and from drought in the same places, with one element of nature's overabundance giving way to another element in some quixotic and perverse display of indifference to man.

At Ballarat, where Australia's labor movement became a force in the nineteenth century, when the Eureka Stockade affair killed many gold miners, I was interviewed by the press and queried about Australian history. I could have said that it was a history of drought and fire but I stuck to the golden view—Ballarat had forced Australia to come of social age, long before American trade unionism became vertical. But Ballarat, founded in 1851, also had many Americans, who went to Australia after the California gold strikes had ended, contributing to the sociological folkways of the antipodes. And Ballarat was to be singularly honored during the Games, for the boat racing was to take place on its Lake Wendouree, two miles away, now sparkling in the undulating countryside. But it was no longer gold but cattle that made 43,000 Ballaratians proud of their social and political history in 1956, the year of the Olympiad.

A month before the Games I finally reached Melbourne, the handsome Victorian city gone athletically cosmopolitan. Nature had put the soft, summery Yarra River almost in the heart of the city; where, to its liquid pennants, oarsmen raced over the river, past colorful decorations and rippling bunting put up for the Games. Melbourne had taken the world unto itself; but she was having delirium tremens, fearing that nothing would be ready. The Olympic Stadium was still undergoing basic renovations; the futurist-looking swimming pool was not even assured of water, let alone completion in time; the cycling track looked like the Eyre Highway after a rain; but all of Australia was cheering, and so was Avery Brundage, the American president of the International Olympic Committee, whose dire forebodings of failure had rattled, riled, then inspired the Australians to achieve monumental records for construction, production and exhaustion, so that all would be completed in time for the opening ceremony. Mr. "Wavery" Avery had been cartooned in the press and almost had his effigy burning on Collins Street; but with due Australian concern for timing in matters international, everything was completed, and the world heard and saw the greatest Games produced in the modern era. It was an awe-inspiring spectacle, as it fanfared and salvoed into human history.

Through the good offices of the P.E.N. Club, I found a room at the Amaroo, a private hotel run by Cay and Bill Morton, who had given up the frontier life of the isolated Northern

Territory, where they ran a station with lots of cattle, for the lesser pastoral life of hotelkeepers in Melbourne. Cay wrote short stories of a sentimental vintage, loved everybody, worked too hard to keep her paying guests happy in her nineteenth-century mansion, and seldom wrote a letter, let alone sad stories about the bush and the outback. But she managed to be P.E.N.'s secretary, giving me tea or beer whenever I opened the door, at which time she would start a conversation about some authors she had promised to read fifteen years ago, when she was frontier bound. The books had not been read, though she still had them neatly stacked, in alphabetical disorder. Before long a novel of mine joined the stack, with her daily reminders over cold beer on a hot afternoon that she would reach me in due time—about thirty light years away.

The Ammaroo was a rendezvous for New Australians, authors in self-exile from Rumania, Hungary and England, who had joined P.E.N. to live in the precincts of literature, something not always evident in Melbourne's intellectual ennui. Melbourne also had no theater to speak of, though hit plays as well as the Sydney Symphony made regular seasonal visits there; but writers, that peculiar breed bred on the typewriter, were thought to be human dingos, the wild dogs that raced in the bush and bit you when they came in packs. But P.E.N., Australia, was more like a lap dog than a dingo, with hardly a bite in its little mouth.

Most of the authors wrote as if they had no teeth at all, though many Australian novels had been on the best-seller list in the United States, with the stigmata of commercialism following one ad behind. Many of the P.E.N.'s members were Sunday writers, budding like amateur painters, giving their work to their friends when they got married or remarried. One author wrote a pamphlet on the social condition of Australian womanhood, insisting that Australia was heading for destruction unless the women armed themselves with something more than a free exchange of libidinous instincts; that the marriage bed was not democratic enough; and to be a career girl a girl had to act as tough as a man, only more so. Other P.E.N. members wrote charming little books of poems, which they distributed to each other. I knew the interior monologue, the Poetry Society of America sunshine that radiated from their collected poetry, the shy, hesitant boldness, which was hardly humility, as they presented their books and patiently inscribed their love and affectations to each other. I prefer humility, which the professionals have, along with some cash.

Some of the good professionals of Melbourne P.E.N. had English and American agents who sold their stories to the *Saturday Evening Post, Colliers* and *John Bull,* though few made writing their full-time hazard. Cyril Pearl, the former editor of the *Sydney Mirror,* wrote flavored period pieces and made a "go of them" in England and the United States. He was inventive, occasionally nonsacred with the fetishes of history, never profane and seldom profound. But he was witty and could tell stories that soon rattled the glass in your hand, as you shook with hysterical whoops of joy. But psychologically, if not physically, literary Melbourne had the atmosphere of the Poetry Society of America: tea, polite talk and little literature.

At the annual P.E.N. dinner, I found myself substituting for Oscar Hammerstein. I repeated my Adelaide speech, though I talked more about American writers being whipped into line by the "need" to pander to both the publisher and the public, with the "horror" and the "pleasures" of contemporary American literature more horrendous than joyous. The facts and figures of solvent publishing shocked my audience, who reasoned that every American publisher was easily a millionaire, especially since television and Hollywood bought so many works of literature. I said that few writers were kept in whiskey or in soup, citing Malcolm Cowley's excellent portrait of the American writer to document the American author's battle for a valid way of life. Five hours later and fifty questions past the time for ending this rapturous conclave, we had bad coffee, served no doubt by waiters who were sensitive, if not inspired, by the sad state of the Australian cuisine and American letters.

Unlike American magazines of mass appeal, the Australian magazines paid about twenty dollars for a short story, and so an author did best to have a rich father or a richer wife or a job as a schoolteacher or in public relations, where many ended up after a few years of literary activity. Others fled away to colder climates, like London's, where they found greater rewards in living their letters in the climate of literature, if with roast beef gone stale and the beer staler. And though Australia had taken in a million migrants, its most creative citizens were still fleeing, unable to await the coming renaissance in the antipodes, preferring the scarred Elysian battlefield of European art and letters with its known sensual appeals, filling all the ships leaving Australia, Europe bound. I had gone through that ten years before, when I thought of colonizing in Australia. I loved the land and its people but not its indifference when books, not beer, challenged the man with the quicksilver humor and the leaden arm.

At Cyril Pearl's one night, my affable, colorful host, who collaborated with his wife, Irma, on his 19th century extravaganzas, said: "Take Bob, for instance." And he mentioned the last name of an Australian writer who had served six months in prison for alleged bad taste in one of his novels and who had fled to Paris to seek freedom for either the good or the bad taste that his style demanded. "Take Bob," said Cyril again. "A few years ago he was wooing a Dutch girl; so was a famous Sydney poet and journalist, but each unknown to the other. Bob saw her for the first part of the week and kept her in true Dutch fashion; the poet in Sydney kept her for the latter part of the week, in even more regal fashion. In due time she promised to marry Bob and she also promised to marry the Sydney poet, with the weddings to take place in a year. Before the year was up she begged to be allowed to visit Holland before getting married; and so each gave her money for a ticket. Bob saw her to the ship and bid her goodbye at Port Melbourne. The ship then went to Sydney, where the Sydney poet gave her another send-off and more money. She had double money, two lovers and double send-offs, though her virtue was slightly not intact, as you can imagine. On the ship she met still another Australian, who came from Adelaide; and he too contributed to her good and welfare fund. When she reached Holland, she was met at the dock by a Dutchman, who married her in a few days; which proves, my dear Harry, that when the bloody Pommies say that we are colonials, we are definitely that, at least; the wild colonial suitors from the bush country."

I knew Bob, who had married a beautiful Syrian and was living in Paris. He kept her locked away, slightly off his romantic style when it came to basic elements of life and living, considering his Dutch experience. He hated Australia and said he would never return, not that he was being invited back to write another "offensive" novel, which had sold very well in the United States, where it offended nobody at all. After all, what was there to love and sex that we had not written about?

But these negative reports did not make Australia; and a country that appears to be inhospitable to creative people can also be deadly for others, in some peculiar exchange of direction. Italy sends abroad its millions, so they can live (though 750,000 Americans go to Italy every summer to worship its art, its pastoral scenery and its antique ways). Thousands of art-loving Italians have settled in Australia; and they live and flourish, giving Melbourne places that make it zestful and even bohemian, especially the Bistro, where one stands up for lunch and listens to the intellectual elite glow in their wine glassses.

When the Suez crisis took conservative Prime Minister Menzies to Egypt, the Labor Party accused him of pro-colonialism. When Menzies returned and gave what most thought was a sober report of his dead-end meetings with Nasser, the Labor opposition accused him of every crime in Australian politics—mainly, still being in power after the "failure" of his mission. It was cat, dog and hyena, after an Australian barroom fashion, with the dignified Prime Minister, who had unfortunately usurped his own Foreign Minister's activities in going to Nasser, roundly abused by his own end of the political arena.

This was only too American as a method of conduct in political affairs; and though Australians insist they are a different sort of breed, hardly of the psychological stuff that Americans are made of, I found the Australian a heady mixture of both the American and the English, but with a third element—his own proud Australianism. For he is, above all, proud, tough and resilient, amiable as a koala bear, yet indifferent to much that is painful, an amazing human being in things that make democracies so baffling to others. If the Englishman is a "bloody Pommy," who appears too stuffed in his shirt and jacket to the average Australian, he is also of kith and kin; and every Australian would think it the last tragedy for them if the Empire were fragmented and the regal emblems done with. Though Australia has much to gain by Empire preference and its dependence on the sterling bloc, Australia has its military security assured by its closer contact with the American dollar. But trade follows the Union Jack as a symbol of political adhesion, even though the Empire keeps losing strength as a economic union of English-speaking peoples around the world.

When England was still Labor, along with New Zealand and Australia, I had hoped that Labor in its Empire dress would show a new road to social progress, to expose the Russian sham; but Labor was soon fragmented, and the vision ended in bitterness. But today Labor may be returned again to challenge communism, though the Labor statesmen are too few. Herbert Vere Evatt, the former United Nations delegate when Australian Labor was in power, has lost his psychological controls, if not his following, by his larrikin acting. He compromised his party by his actions during the Petrov spy case, when he argued as if he were on a soapbox, for cheap applause. But Labor is split and Labor is also Catholic when it is not something else to the Left of Christ in overalls—and though the communists are not electorally strong in Australia, they control many unions, including the dock workers and the miners, or the life blood of Australian industry and commerce. Recent agreements have tended to cut down the strikes of the Wharfies, who often walked off a job because the tea was served cold during a coffee break.

The Games fever soon removed politics; and I often scootered to Heidelberg, to Olympic Village, to watch the athletes train or to get an interview. A friend of mine, Jane Williams, who lived opposite the Olympic Stadium, offered me her apartment for the duration of the Games, saying, "I hate cricket, football and track, though I ran at college; and above all, I hate mobs going to games. Harry, please move in for two weeks while I move out of town."

I moved in and had an emu's eye view of the stadium and the dry runs. The excitement had everybody on edge or over the edge. Parties, balls, dances, and foreign visitors were the order of the day. Ladies of easy virtue became easier as they stopped strangers on St. Kilda Road and offered them Olympian sports of another sort. Prices went higher than the sky as hotels tripled their rent. Little scandals became bigger and bigger scandals soon reached their disappearing proportions . . . but above all there were people dancing at balls, waiting for November 22, when the Games were officially opened.

I went to the Lord Mayor's Ball at Town Hall, arriving in a tuxedo on my scooter. Not having black shoes, I wore dark brown, which were hardly noticed until a girl, from too much love of consuming, soon found herself under the table. She was tapping my shoes and announcing, "Harry, I must be going color blind. Look, everybody, Harry's black shoes are changing pigment." After that, and since the Lord Mayor's Ball had reached that state of fancy, I wore my brown shoes around my neck, and I was not the only one.

The Games were on and so was the Hungarian revolution. Some of the Hungarian athletes had been on the barricades before they rediscovered they were sportsmen above all. When the International Rescue Committee cabled me to find out how many would defect, I found myself doing research of another sort: How many Hungarian athletes wanted to defect to Canada, Australia and the United States? After every visit to Olympic Village, the number would change, going from fifty to seventy-five; then, after a night of self-torture, only a handful said they would defect to the democratic countries. When the International Rescue Committee sent me money to distribute to them, it took hours of convincing to make the forty-seven defecting athletes put out their hands for the small checks. We were also in the middle of a money-making drama, with many unofficial spokesmen for the American A.A.U. acting as busy little plenipotentiaries of profit, trying to turn the famous Hungarian water-polo team, who had defected en masse, into professionals.

But it was a bitter drama all the way, especially when the planes took the athletes back to Hungary. Many intended to get off in Vienna; others, uncertain, were numbed with fear as the local Hungarians at the Melbourne airport begged them to get off now—that only death awaited them in Hungary. I cried, along with others. The Olympiad was over and the victors were the losers. We in the West had once again failed to exploit the revolution against communism.

With the Games over, I took Olympian stock of my total financial and literary assets. I had written four poems, all told, since leaving New York a year before. I was no longer the "wild" poet who wrote a dozen a week; actually it became a minor sort of subjective celebration if I wrote one a month. The source was not drying up, for I still made mental notes of ideas I intended to use when I could manage to get financially ahead of myself. I had always worked nine months of the year at commercial writing, giving birth to novels and articles under pseudonyms, with every fourth article one I did not mind putting my name to. Years before, I had worked out a peculiar mathematics to earn my living by writing for an assortment of men's magazines. I usually did a story a week, at home, as well as several pot-boiler novels during the nine months that I devoted to the economics of bread without butter; and in the three months left, I wrote a serious novel as well as poetry, hoping to fill in the hollowed-out man, who might well be losing his sensitive tendrils between Grub Street and Grab Boulevard.

If the poet had failed himself, the gentleman of purple prose had not. I had written an article a week since going to Europe, with the tremblings of the scooter giving my thinking and typing an odder rhythm than was my normal writing style. The book I was writing of my odyssey on two wheels brought out the still-present poet; but in the place of poems I wrote about the folkways of places seen and people observed. Before leaving Paris I had done research articles on the Folies-Bergère and Place Pigalle, on the Lapin Agile and the Butte on the Right Bank. But articles on Paris and its different sorts of nights hardly filled my purse to the breaking point, though I had enough to get me the scooter and give me at least two months of gypsying over ancient unpaved roads that Genghis Khan might have ridden. I had also done an article on the famous pubs of England, with Henry Treece in Lincolnshire for an able helpmate, as we pub-crawled from Lincolnshire to Nottingham in his Woolsley car. Another article that I enjoyed doing was a piece about life in a Turkish village written for a most conservative magazine, one that you can safely be seen dead with. But from the sublime to the conservative magazine is only a matter of so much more a word, and when my words were good, I was happy.

My checking account back in New York was not too liquid with dollars; and when things became overdramatic in my overdrafts, I cabled my brothers to set things to rights. As a matter of fact, they cabled money to Teheran, which arrived while I was deep in the heart of dysentery, in Omar Khayyam's Khurasan world in Nishapur. I was assured that Bank Melli in Teheran would apply this money to my New York account. Believing this, I drew several small checks, convinced that international banking can do no wrong in time and place when it comes to the simple transference of currency. But Vice-Consul Charles York, an aide to matters American at the American consulate in Melbourne, wrote me how wrong I was. My three checks had bounced from Iran to India to Australia and back to New York, in an involved financial boomeranging. But it was the season of the good heart, and the bouncing money soon bounced back, when the brotherly checks showed up again and were applied to my New York account.

My room at the Ammaroo, to which I returned after being a next-door neighbor to the Olympic Stadium, was a large nineteenth-century drawing room, with carved bookcases and huge mirrors which were everywhere but on the ceiling, where I expected them as well. Prior to the Mortons' taking over this strangely magnificent building, it had been the home of wealthy Italians who had bought it from some unknown people, known well enough to Melbourne's leading sporting characters for running the handsomest bordello in town. But the neighborhood became too spiritual when a synagogue opened directly across from its nonsacred portals, and the ladies were spirited away and bedded down elsewhere. My room doubtless had been quite a center for horse racing, by its size alone. Outside my room, the curving staircase must have made the ladies look nudely decorous as they mounted it for things less horsey, on their sporting rounds.

I received all sorts of gifts before my departure, and I gave of the only thing that I had to give—myself. An ex-Marxist poet, turned Buddhist priest, gave me a privately printed book of Chinese poetry that he had most privately translated and then printed on his proofing press. The Buddhist priest turned up every Monday at the Mortons, ate longly, for he was always broke, then blessed everybody in sight. During the Games he was the high priest for all the Oriental teams without a holy man to help them, and my friend cycled out in his robes, sans beard but filled with Buddhist joy, to help the Thailandese and others who practised Hinayana Buddhism on to athletic victory.

The Hungarian community gave me a translation of Hungarian poetry; the owner of the Bistro, a handmade, privately scrolled menu featuring *Roulage with Sauce Roskolenko*. It was a peppery dish, though it looked like an abstract painting by a girl I knew. The girl was an extremely good artist, fey and extraordinarily beautiful, more like a Modigliani painting, with an elongated neck that made her head and naked shoulders a thing of joy. She supported herself by teaching abstract art in a lunatic asylum, where the abstracted art of living had probably brought the inmates originally. She gave me an excellently conceived painting, "Abstraction in Red," when I left lyrical Melbourne via a taxi for Essendon Airport. I lugged the large canvas, acutely appreciative of her many-sided talents and her loveliness in a worldly world, where tragedy has its romantic kin at every juncture. Her companionship, then, made me forgetful and hard-working, and I did not often remember the reasons for my own flight from America—the divorce that had driven me around the world. She is or was an amazing thing of poetry, in all her ways.

The American Olympic Association had given me a handsome Fact Book, but the facts were only too sad after the Games, especially for those who denigrate the American Negro. Without the Negroes, we would have trailed the Russians, if not the third-place Australians, in track and field. Our white women looked better off the playing fields, when not competing; and a few Negro girls brought in our only female wins or placings. One-tenth of our nation brought in half of our athletic victories, which should be extraordinary facts and figures to fulminating Southern politicians.

Odd-and-end arts were given me—tokens of my return to Melbourne after being away for ten years. But I was leaving again, and once more I paid back for the twenty-eight dinners and parties that I had been invited to during my Melbourne stay. I borrowed the kitchen, the garden and the whole of Mitta and Alex Hamilton's old home near the Yarra river; and with Maria Kozlik, the cooking editor of the *Melbourne Age,* for my assistant, I cooked for twenty-eight friends. Later, Maria Kozlik sent me a clipping of the *Melbourne Age,* which said, in part: "In a moonlit garden near the Yarra, I watched Harry cook a Chinese dinner for all of Melbourne. While turning skillfully the shredded beef, which sizzled in the frying pan, he discoursed on topics ranging from the tragedy of Dylan Thomas to Australian art, from Japanese sukiyaki to the paintings of Fujita, and about wine from South Australia, which he said was as good as the French."

I had said good-by to Melbourne so often that I wondered what it was that brought me back. Was it old friends, a sense of living over the war years, when Australia had been home base between trips to New Guinea and the Philippines; or the fancies easing me into middle age? Australia's frontier had always engaged my imagination; but at forty-nine, my talents for frontiering were slightly negative. It must have been the Australian personality; though I'd had enough of their easy ways, especially their delaying tactics, which they apparently took on loan from the Mexicans, preferring to postpone until the year 2158 what they knew they would not get by 2058. With this sort of time-space-progress mathematics, it made Australia at least twenty years behind the United States in all things good and bad in gadgetry civilization. But I was leaving for Sydney, then home on the S.S. Orsova five days later, the sentimental man who had found Australia wonderful in all man's ways.

Sydney was old-home week, with acute memories of Sydney 1943, 1944, 1945, when I lived, sinfully, with a blonde lady poet between chores of war and not dying up in New Guinea's muddy ooze. For we who were not about to die saluted the senses of love with keen resilience and bounce. But the blonde lady poet had gone, though once it had almost been marriage. We had planned it but it had not taken place because one of us could not make two of us going down the aisle. She had changed her poetic mind. But as I walked toward blue, sail-filled Elizabeth Bay, I wondered at the constancy of some and the relaxed morality of others. What made some people dull and moral, others dull and amoral, and the third group an exciting admixture, no matter the nature of their conservatism? It was the person and the place and the echo of some warmth, which religion unfortunately no longer gives.

My past had been intensely religious, though I had strayed mightily when I grew up. But I was still the spiritual-minded man no matter whether I was in a church, a mosque, a synagogue or a Buddhist temple. My poetry, however, had given off Pantheist echoes, which Henry Treece had called "apocalyptic" and which Lewis Mumford had called "my thunderous days." For I was filled with wrath against social iniquities; and what Gorham Munson had once called "my acute self-involvement" ended up with Paul Rosenfeld's statement, "your poetry is grand, severe, bitter but powerful."

It was a social religion, making some men Liberals and others Marxists. We had God hidden, nevertheless, under a variety of clichés dealing with trade unionism, a third party, and an economy of abundance for all. But after the war I shed much of my social attitudes, expecting peace within the world, if only because the United Nations always talked peace. I had seen what the atom bomb could do when I had gone on a journalists' tour of Hiroshima in 1947, and later, in Indochina during the early days of the civil war there. Actually, I had shed very little, I discovered. I was still a good anti-colonial. If the West lost its anti-communist appeals in the colonies, we would lose because of the colonial questions plaguing us.

A few days later I was crossing the blowsy Tasman Sea on the S.S. Orsova, which carried tourists and pilgrims from the sixteenth Olympiad, as well as Australians forced to take the long journey, via San Francisco, to get to England. The Suez was still not working as a watery ditch; and Nasser got his share of "bloodies" from the Australian passengers, whose fare had gone up by a third as a result of Nasser's adventures as a ship scuttler. And the S.S. Orsova was like Rimbaud's poem, "The Drunken Boat," as it floated between green gulfs and intoxicating waters toward its first stop—Auckland.

Chapter 10

ISLANDS IN PARADISE

ON THE S.S. Orsova were half a dozen professional American football players who had tried Australian Rules football in Sydney. They were returning as chastened athletes, their crew-cut heads showing dandruff as well as the scars of combat. They had but one aim, to sleep with all the women passengers they could induce to seduction. One of the lesser idealistic athletes even kept a small book, and daily he regaled his libidinous male friends with his box-score. When one of the inducted girls heard of his muscular bragging, she spilled a glass of beer in his face, saying, "I should have given it to you glass and all, you bloody bhar-stud!"

There were also New Australians migrating to Canada and the United States, having tried and failed to make the summit of Australian frontier values. One of them, a Hungarian woman, had won many prizes as a champion typist, which she proved by typing as a champion for me, doing over a few articles I wanted to sell en route across the Pacific at the four call-in stops we were about to make. I was still the journeyman journalist, coming home after fifteen months of self-imposed exile, drifting through a tropical world that balanced well between the waters and the sky but left some emotional driftwood floating about in the Roskolenko Seas. But memories of bitterness go badly with tropical waters, and as I listened to champion Mrs. E. I.'s many complaints about Australia, I wondered what happened to such people when they lost the edge of their bitterness and found love instead. Likely they would be completely lost away from the acids in their hearts, unable to live with themselves. Had I become one of those? Or had the "malices of divorce" finally washed away? It had not with Mrs. E. I., who said, over coffee in the lounge: "I got tired of Australian mutton, of steak and eggs, of bad coffee, of swimming and sunning, of drinking myself into a stupor. And that is all there was in Melbourne for me. In New York I will find a new husband and raise a family, as my Budapest husband seems to have divorced me since I ran away from communist Hungary. He said that he would follow me; but he followed some girl communist into the cause and has become a bloody communist, too. My life, however, has become a peculiar goulash—all spice and no meat."

For a Hungarian woman, it was an odd revelation, making her a dubious candidate for the spice islands of the crew-cut crew. I was a candidate for her typing, and each day she did five pages, the sad champion lost in the limbo of a husbandless life. The Hungarian anti-communist revolution had, however, brought a letter from her ex-husband—could she send him money to buy a suit, including the necessary underthings? She couldn't, or wouldn't, remarking, "He can go about in the nude, if he's that idealistic about his new wife. I suppose I should send her a gift, like nylons or panties or a girdle or a letter on how to sleep with my ex-husband or what he likes to eat. He likes to eat," and the conversation halted. The waiter had come around with ice cream and a mob of passengers pounced on the waiter, though they had just finished lunch.

"Everybody likes to eat," she went on as another passenger joined us. Tony, a half-Indian from San Francisco, had taken his wife and two young children to Australia ". . . To go bush, to work hard, to make wampum, and to see what a real frontier, which I never knew on the Golden Gate, 'cause I was born too late, looked like."

As he ate his ice cream, his thin, copper-hued face-muscles were working angrily, as if the ice cream were the antithesis of a kangaroo steak on the Australian frontier. "And the frontier looked like hell!" he said, exploding into four-letter words, all of them censorable. He apologized to the Hungarian woman, who laughed off his anger, and added, "I can do better than that, but I don't want to shock Mr. Roskolenko, who thinks that I am a gracious lady."

"Please shock me," I said.

Tony had been a partner in a small, defunct lumber mill; but his Australian partner forced him out after Tony had made the mill profitable. On reaching San Francisco, Tony wrote a letter about his lumbering experiences in Australia and sent it to a paper in Oakland. The Australian consul in San Francisco answered it, and soon it was a free-for-all, with the consul getting the better of the mephitic argument about lumber, prices and partners.

I was almost broke by the time we reached Suva in the Fiji Islands, where less than a century ago money was an oddity of no value; but modern Suva was a busy commercial outpost in the Pacific. It even had a daily paper, and so I wrote a kaleidoscopic piece about my travels, which I took to the Irish editor of the *Fiji Times*. He bought it, paid me, took me on a tour of his printing plant and later introduced me to the Consul General of India, who toured me for the rest of the day through the idyllic background, amid the coconut trees and palmy villages. The new Fiji was in the midst of catching up with the past as the natives, whittling away at arts and crafts, waited for the passenger ships to call in and twentieth century passengers to traipse ashore and bundle back with carved knickknacks of a lost, splendid past.

The handsome Fijians, wearing their hair in towering coiffures, were impressive, childish and kind, but without too much sacred love for the Indians, who outnumbered the Fijians now by more than 30,000. But commerce can count with ease and multiplies easier; and since Fiji was the center of the Polynesian grouping of islands, midway between the areas of paradise, it was no longer lost, having been discovered as a middleman's sacred source of contentment. All these islands had Chinese and Indians, who had left their million-headed heartlands for the Pacific wastes, where they cultivated golden crops in the colorful Oceania they now inhabited. It was another world but much the same commerce; a gossipy, palmy midway to the banks. It had lost its real Fijian arts and given the Fijians bicycles, along with oil stoves, to streamline the places of paradise.

Several hundred miles away was Samoa, where I had once lived briefly. I recalled Mauri, a lovely eighteen-year-old Polynesian girl I had met there ten years before. She was walking through the village of Apia, her head high, her body uncomplex, her instincts simple; and she was singing a Samoan song, gentle with lilting vowels and cadences. She wore a red wraparound that exposed her conical breasts, completely indifferent to my Western-staring eyes, though the indifference did not last. In too short a time, she invited me to her home. Soon we were wandering through a deep woods, past the Vail of Vailima. There, she pointed upward, toward Mount Vaai, saying, "Mr. Robert Louis Stevenson is up there. He is buried under a big stone that says, 'Home is the sailor, home from the sea.' We must climb it, sir."

We climbed it; I, with much puffing; she, with ease. It took almost an hour to make the steep, thousand-foot ascent. In the dark the large black tombstone appeared to embrace all of Apia, including the ocean and the beach, where wading fishermen, holding flares aloft, were singing as they fished.

Trembling, I suddenly associated fleeting bits about the story-teller, Tusitala; Robert Louis Stevenson, the man and artist now buried beneath the stone; the exile from the city looking for health, who had finally found it here at the Vail of Vailima, where his old house had been taken over by the government. He had helped to stop a civil war among the Samoans, something the warships of England, Germany and the United States could not do. One of the ships was still in the harbor, having been blown to bits by a storm.

In the presence of life, or within the dark eminence of the tombstone covering Tusitala, I fumbled with the words of a poem, scrawling it on an envelope in the blackness:

Through green corrugations, the trail's tall ascent
Rises from Vailima, where his house still stands;
Fifty years ago he died in the valley,
Where are the boys who mourned him then?

They carried the author to the little high mountain,
Built a grave and buried him there;
The trees stand mourning, they weep and howl,
The winds blow straightly above his mound.

They carved his name into the stone,
Home is he who a home had never;
This is the island little children treasure,
Home is the sailor beneath this stone.

Mauri, I remembered, has asked me to read it. I did, wondering if it made any sense to her; though my last line, a variant on Stevenson's, she had quoted correctly as we were nearing the Vail of Vailima. But then, R.L.S. was basic education in Samoa; the poet and adventurer was root culture here. But what was not part of my basic education was her sudden proposal that I give her a "blue-eyed baby."

I had lived there too briefly. Had I stayed longer, I would have missed my marriage to a Eurasian and gone Polynesian instead. It would have been less complex, emotionally, stabilized on a Pacific reef, but one that Robert Louis Stevenson had finally found. I was surrounded by the nineteenth century literary ghosts that R.L.S., the engineer turned lawyer turned poet-novelist, had created as the story-teller of Polynesia. Stevenson had been the wanderer seeking a haven for his health; but I was only an emotional free-booter, suddenly asked to sponsor a possible blue-eyed son by Mauri. The racial irony, if not the honor, kept me from becoming a father by proxy.

I was tempted to revisit Samoa to see Mauri. Instead, I accepted Fiji as a sufficient token of cheap fruitfulness; for it was the same, if a more crowded island in an ocean of no escape, with nature still acting bountifully. Or so said shopkeeper-poet, Laurence Dakin, the Canadian writer who had come to Fiji for a brief visit, married a beautiful Polynesian girl, raised a child and been idyllic now for five mellow years. He owned an arts and crafts shop, paying his way in the Fijian paradise when the tourists came ashore for twelve furtive hours of going primitive. Between rich tourists and the Fijian lack of furies, Dakin wrote poetry about Fijian myths, as well as literary studies of Ernest Dowson and other romantic writers. But he wanted Europe, desperately; and he wrote me, later, that he was planning to go to Italy.

There was real gold in Fiji and four thousand natives tore it from the earth on the north coast at Viti Levu; but there was more than gold in the lush gardens and the byways. The radiant Fijian, from his thatched hut, looked upon the white man, if not the Indian, as his locomotive into Western history. I got a two-hour view of the social setting, a pamphlet of poems from Laurence Dakin, a calendar of Indian holidays from the pleasant high commissioner, a large rose from his young daughter, and some trinkets of the past that is so long past that it is impossible to contemplate the commercial carvings the natives sell without cursing the white man's missionary stupidities and explorations that have forced these primitives onto the Western locomotive, derailing their present and trinketing their past. An ashtray that Dakin also gave me showed a Fijian all garlanded with flowers. On the back, it said, "Made in England."

It was sea adventurers and explorers like Abel Tasman and Captain James Cook who had sailed into this primitive world and opened it to the caravels of commerce. Captain Bligh had come by in 1789, preparing the bounties for Hollywood. They were the early self-exiles, before expatriatism became a guide to aesthetic and political values. But the closer I came to home, the more realistic was the view of my own flight. I was not an eccentric adventurer, but a modern man who had left the shambles of his private life for a self-contained exile, hoping, en route, to work through the emotional havoc. I had come through, self-purged, and the better man for it.

The twenty-two days at sea gave me a lot of time to examine my past. It was always present, just as the ocean was; like the subteraqueous tides making the surface of the ocean swirl, bringing up seaweed. Later, when my objectivity found some order, I thought of my fellow-passengers: men without lyrical scooters; women without dreams; people with and without the wherewithal to carry on, yet carrying on in strange sequences and places, as if some great touchstone was at work; and now, with the magical balance, they lived their new lives in the tropics, preferring to ride it out in their middle years as they remembered a youth they never really enjoyed. Now they were the uninhibited citizens of the Polynesian world, on the watery frontiers that Tasman and Cook had place-named for them. Yet they were confused between segments of the past. Their new vista, which they accepted physically, came into conflict with their mental habits and Western-remembered tensions. The modern primitives were not primitive enough to escape into the escape they so desperately wanted.

Honolulu is America's Polynesian paradise and the soft underbelly of the Eurasian world. Here the great mingling of the races takes place: Japanese, Chinese, Polynesian, Negrito and White live in a complex order of private and public worlds and sentiments. They intermarry, after a fashion, and bring forth the new breed, a much handsomer type of human being. But Honolulu is essentially sun, beach and escape—a vacationland for people seeking the limbo latitudes of the sun. I sought some cash, however, for I was almost broke.

Once again I sold the article "Kaleidoscope," with the editor of the *Star-Bulletin* throwing in a drink over his cluttered desk as he read the paying-off piece. He flattered me, took a photo of mine showing pilgrims from Afghanistan on their way back to holy Meshed, then said, "Iran is like parts of New Mexico, isn't it? What about the ancient Zoroastrian religion—is it still practiced?"

I had been indifferent to the once heady fumes of Neitzsche's *Zarathustra* while in Iran, though I had read him avidly as a boy. As for the ancient Zoroastrian religion, it had had its throat slit, down to the crotch, by government decree. The Shia Moslem religion was the state religion of Iran, once the birthplace of the Zarathustrian concept of good and evil or the battle between the forces of light and darkness.

We talked about D. H. Lawrence and New Mexico, about

sun-worshippers and the Penitentes, about poets and Neitzsche, who had once underscored the relationship between poets and the truth. Somewhere in *Thus Spake Zarathustra,* Neitzsche had written: "But granted that someone did say that the poets lie too much: he was right—we do lie too much. We also know too little, and are bad learners: so we are obliged to lie. And which of us hath not adulterated his wine?" I had too often.

As for D. H. Lawrence, the myth about him was greater than the man and his works. He was the Zarathustra of Taos, New Mexico; the godhead of natural living; the man in the sun. He was also, unconsciously, the father of sex in the modern novel. He had helped to release the Puritanical American language; and now a paperback glorification of four-letter literature was all over the land.

As for the smut of the sexual-minded novel, I had made my own little contribution toward its latter-day phase. I had lived within its purpling libido world for five years, writing fifteen sexy books for the cash it brought, about a thousand dollars per contribution, including the small royalties the publishers of these teasing titles gave to their hungry authors. But each book took its literary toll, erasing the poet's simpler decencies and adding to the lie. I had become the nonwriting poet during those five years. It was easier for the conscience, oddly, not to write poetry while I was engaged in writing libidinous novels, which had more customers. Often I took three weeks to write a book; but with a long day and night at the typewriter, it was more like six weeks of concentration. Once I wrote around the clock to make a deadline, discovering in the early hours of morning that I could get some extra-sensitive deception if I wrote dialogue that went, almost one word per line, like this:

"Lay?"
"Today?"
"Now."
"Okay!"
"Any good?"
"Perhaps."
"How good?"

"How now . . ."
"Like loving?"
"Love loving."
"Love everything?"
"Of course!"

And so forth for twenty-one pages, with rises and falls of intensity and direction shunting between the languishings of sex and the language of the paperback ideal. My books paid some bills.

The editor of the *Star-Bulletin* had come from the mainland, the land of enterprise, ulcers and tensions. But Honolulu softened a man; the underbelly gave in when poked with a finger. One vegetated as one drifted under the Hawaiian sun, away from the mainland challenges. But here one found nature; some physical, tropical inducements; a kind of simple purity. In Iran the purifying element had once been Ahura Mazdah, the Persian Messiah. It had been the wide spaces and the worship of agriculture along with the sun. But the Persians had at least created Persepolis—a marble deification of sun worship, good and evil, light and darkness. In Honolulu there were only neon-flashing tourist traps, deifying commerce and not religious, messianic places. But all religions are closely connected to spiritual larceny, borrowing or stealing from each other. I knew that poets also had larceny in their workings, but as a much more delicate art. When the editor handed me a check for my article that had by now been much printed and reprinted, I accepted it along with a second drink and went my larcenous way.

I was returning home to my New York womb, with the sea voyage merely an interlude before I faced myself again on the brick fastness. By the time we reached Vancouver I had analyzed the excesses of my hazardous journey around the world, as well as the intimate voyages I had taken within myself. The joys of sudden kinship in some strange place and the varied misadventures symbolized the points of tolerance in personal freedom; the excesses, the strains of personality feeling for a direction. But Vancouver was North American soil, with its Canadian similarities somewhat roughened up; the city leading to the frontier, to which many migrants, tired of Australia, were going to make a home. But "home" was the endless token and symbol; home that did not have unmanageable ghosts running through the furnishings. I had ghosts everywhere now; for in each country I had left part of myself and taken along part of another person—the baggage of other wanderers on the lowroads.

I remembered Vancouver, having been partially responsible for sending George and Inger Woodcock there from London, eight years earlier, when George lamented about England's future disappearance as a world to live in any more. George was a philosophical anarchist, or he had been, once writing serious books about historians, economists and anarchistic philosophers who had played with socio-political ideas to help release or remove the excesses of capitalistic politics. George, now much removed from the passion for freedom, taught at Vancouver University; but between scholastic chores, he and Inger went abroad for a year at a time to refuel on Europe's past so as to make their present, in Vancouver, tolerable or possible. Inger, a German potter, was filled with zest and energy; and one could always gossip about the folkways of the energetic friends we had in common, those who had saved themselves handsomely while not saving the world. At fifty the folkways became folklier; the splendid self adrift in anarchism, less splendid; for middle age can only talk into a mug of beer and get back a foamy answer. We sat in the tap-room of a Vancouver hotel and talked about the lives and deaths of friends; about Stephen Schimanski, who had flown off to cover the Korean war and flown into the sea, en route to the fighting; of Henry Treece, who preferred to write about King Arthur's times and troubles, to our own times' challenges; about the *Partisan Review,* which had whored after success and found only a faint echo of something quite vague; of Dylan Thomas, who had never been anarchist enough, though privately, poetically disciplined, but mostly so that he could die before forty; of our own little battles and what we had found after the field had been cleared of the ravages we ourselves had committed in the name of idealism.

Chapter 11

THE ROAD HOME

EVERY MOVE of mine was now a visit to my past; in every city one found a friend somewhat battle-worn from past socio-political engagements, yet always ready to talk over the world we never remade. Two days later, when the S.S. Orsova passed under the Golden Gate Bridge and anchored at the Embarcadero, in America's only Mediterranean city, I knew that I was home; that soon I must scooter across the United States, once again on two wheels, after many months of sitting out my odyssey in Australia.

I had lived in San Francisco on several occasions, beginning in 1928, when I interrupted a hoboing trip across the United States, using a freight train from which to see America first. It was the magnificent city glowing beneath the usual diamonds of the fog, a city that one left sadly but that one returned to happily, a city strange, uncluttered, rich in its hilly, Midi frames. It was Mediterranean even to its earthquakes, a city trembling in its peculiar relationship to the rest of the country. One looked, one accepted, and then one lived on a hill in San Francisco.

It had many war meanings for me; the hurried romance that went with guns. In 1943 I had studied celestial navigation there, prior to taking off for places known and unknown, though I eventually landed on a New Guinea beachhead. I piloted small ships during the war . . . but in San Francisco, I "not about to die," piloted myself to a lovely lady's bed; everything quickened by death's peculiar relationship to love. Elizabeth loved love, she said; and Elizabeth loved me, she said; and I loved her, I said, both of us believing both of us.

She had a flat on Union Street, just above the depths of the Embarcadero, and every morning I would watch the incoming transports. One morning I would be a passenger in uniform, outward bound for some Pacific area of dread. In between, I studied cosmic things about the stars and we listened to old English music, especially to Purcell's *Dido and Aeneas*. When I went off to the wars, finally, I took the first recording of the opera in my bulking foot-locker and I was to play it often in the New Guinea jungles, hardly the sultry setting for a lamenting Dido about to commit suicide over Aeneas of Carthage or the barbarian King, Iarbas. But no matter the doubtful legend, the music was endlessly fascinating in the flat on Union Street. Besides, Elizabeth preferred fornication to suicide. We were about to become war-torn lovers, dying daily in the hundreds of letters that we wrote to each other for several years afterward. But what a beachhead that was until one night, six weeks later, I said as I looked out over the Embarcadero, "It's that one, I think." She cried—and I was on it, twelve hours later, and away.

When I returned from the war, filled with malaria and sporting a broken shoulder, I stayed with Kenneth Rexroth for ten days, rediscovering civilian strangeness, the normal man again after a few years of military murder. And now I was once again in Rexroth's book-lined home, back from my private battles, and about to discover the regional renaissance in poetry that Rexroth had fermented in the sub-cellars and night clubs, with jazz bands accompanying harsh, unpoetic voices reading the New Realism from a garbage-can reality they were exhuming. It was so real that a *Life* magazine photographer was haunting Rexroth's life of anarchism and art as he organized the clangor, drums and cymbals. He was joined by thin, talented Lawrence Ferlinghetti, an exile from Yonkers and Paris, a poet who loved the bare, barren statement of sociological fact which

William Carlos Williams had made so tremblingly popular for the dead ears of so many contemporary, talentless typists. But if whimsical Williams was the lower priest in this hierarchy of literary anarchy, whine and woe, his San Francisco followers, to equal things out, farted brightly with a flatulence that would have amazed some Elizabethan scholars of the art. But it was hardly inventive poetry despite its basic shock values, its reckless cynicism regarding the social and political scene, its variegated anarchism and cream-puff morality. At best, it was juvenile rebellion jammed to the teeth with slap-dash, beat-generation parlance, the myths of a cool language rendering their disorderly ooze. As for the real poets among this new breed of radicals without a program—they would eventually emerge from their ridiculous attitudinizing, providing they accumulated some moral guts when they exited from their nihilistic stupor and circus. This was the latter-day renaissance, with Rexroth, who is an excellent showman, presenting the jazz-poetry gladiators in the nightclubs along the twilit shores of the Barbary Coast.

A magazine called *Ark,* skillfully edited by James Harmon, published some of the better beat poets; though many of the contributors, especially Richard Eberhart and Louis Zukofsky, came from my generation of non-beat, non-flatulent citizens with a cause. But *Ark,* from San Francisco's geographical and spiritual peaks, was chiefly concerned with disenchanting its readers, taking them down the dusty cellar-road to *beatism.* A few years ago the symbols that equated rebellion had been sociological and political, when *alienation* was the last catch-word for the more astute critics of the American century. But this was San Francisco's renaissance, with *beatism* and *disengagement* sounding the tom-toms of the poets' savage appeals to un-reason. The better poets like Gary Snyder, Philip Wallick, Mitchell Lifton, Cid Corman and the now notorious Allen Ginsberg and Jack Kerouac, had all the surface conceits of talent; but talent was not enough. Kerouac, who had not yet published his novel *On The Road,* was the acknowledged high priest, after Rexroth, of the cult; soon he was to symbolize the depths of the hallucinatory hipster night with his crude, semi-literary language as he reeled, in book-length abandon, from self pity to moral anarchy.

Peculiarly, *The Nation,* the safety valve for things "Leftist" and "Liberal," acted as literary midwife for these barbaric bards of the dungeons. Even *Mademoiselle,* the chic-all of collegiate teenagery, gave up its virgin-like aspects in saluting the butcher boys of poetry. Some observers found neither vigor nor masculinity—just sound effects braying at artificial moons. One perceptive critic, Robert Stock, said: "And it is precisely those modes of the twenties that have dated much of that which most astonishes today's San Franciscans. Kenneth Rexroth may be craftier, but Ferlinghetti is more buoyant and musical as he romps through the rubble of civilization." And what a rubble they were romping through! Allen Ginsberg, who had experienced everything, yet nothing, as if he were another hallucinated Rimbaud but without the poetry, wrote the epic of this bastardized medium. Called "Howl," it did just that, reaching the summit of the faded four-letter phase, and without the real savagery to make it more than clinically interesting. What San Francisco needed only Rexroth could prescribe as the good literary handyman-doctor; though a gigantic vacuum cleaner and kingsize band-aids might have helped stop the seepage of the outhouse going over the non-lyrical structure. It was not Tristam Tzara's *Dada,* nor good comedy nor just bad poetry. It was the off-beat junk of the mind given over to shock and cheap sensations; and high seriousness, which is poetry's first commandment, was hardly visible in most of these guests in Rexroth's genial revolutionary house.

It was organized for bohemian bedlam and without the dancing girls. One cute advertisement to the noisy exhibitionistic sacrifice went like this:

>   NEW SOUNDS—NEW WORDS
>   FERLINGHETTI & REXROTH
>   reading their poems
>   LIPPINCOTT & CELLAR QUARTETTE,
>   making music
> both together all at once simultaneously
> I mean, like man, make it.

Like it will be a real cool scene, like TERRIFIC
Like tell every cat you know. Like bring a chick.
Like bring two chicks.
  THE CELLAR   576 Green
  WED. NITE   Feb. 13

 In the twenties this infantile play-acting would have been somewhat acceptable. But this was a quarter of a century later, and a war later, with much of the world trapped by communism's storm-trooping materialism. I was tired of the critical fraud that the socio-literary parvenus had pulled on us. We were endlessly told that we had nothing but production charts, though every country was trying to out-Americanize us. The Russians were bragging that soon they would overtake us in commodity production, if not in the destruction of simple human values. We were told, and I had done my share of the telling, that only Madison Avenue was talking our culture like so many public relations handouts. But we were many other things, too; and we had a surprising greatness in our midst, though the midst was a Babel of strange silences.

 Only these barbaric echoes, sounding so chic to questionable editors, were the American voices, the American colors, the American depths and the American vices. Our vices, such as they were, now included opium. Once it had been the shimmy, so that even vices tend to change in emphasis. Our vice was the home that was not a home; the fathers that were not fathers; the mothers who were mothers too much. I had been brought up on the strap, which, when well placed at the right time, can stop a howling infant from maturing into a murderous teenager. But especially poets, charmed by their excesses of mental freedom and creative compulsion, need discipline. And in San Francisco they cared for nothing in their poems, blaming their fathers, mothers and even their psychiatrists, for their wheedling nastiness.

 I know that most poets are, intuitively, a sad lot; that they have a well-developed radar for the privately sensitive response, when they are not politically dangerous. It was the critic Lionel Abel who once said that poets are a would-be criminal type, but that their crimes are purged through poetry; and it was a Greek philosopher, or any number of ancient Katzimbalises, who said he would exile all poets, but with wreathes of roses dangling from their heads. But I was an exile in their midst and without the roses. They were strangers talking a language that was all dangling nerves; too brittle instead of brutally sensitive. They could neither objectify nor experience the real any more. At least Rexroth and I had gone through the challenges of the twenties and the thirties, and we had known the birth and death of social and political ideas. We had known jails, beatings, starvation—all the brutalities of growth without "security." And we knew that one did not wail and howl because 1957, in its cosmic coma and political vacuum, was supposedly the end of the world. We had been the infants of the World War I; but we had not blamed our future lives and our ways on that. Infantilism, today, was a dirty child fouling itself, not the self-appointed caretaker of its age. But we, modestly, had been healthy in our arrogance and our revolution against the stultifying agencies and the political strait-jacket of conformism.

No, the world had not died despite Eliot's bang-whimper *coup de grâce* poetry. And whatever one cared to call it, from the Age of Arteriosclerosis in Arts to the Age of Insecurity, it had been quite secure for these infants now plotting their juvenile senility. I listened to the renaissance riot over San Francisco, then I listened to the two lovely Rexroth children, Mary and Catherine, and I found them wonderfully zestful as they mimicked everything they saw and heard in Rexroth's literary pageant. The children, at their own game of mumbling prosody, used dolls, a long staircase, Japanese costumes, the domestic situation, *et al*. They danced *Noa* dramas and uttered gutteral sounds, being completely renaissance in spirit. Mary, who was six, was really talented; whereas little sister Catherine, aged two, had not fully developed her arts; but she had a smile that was worth, per inch, all the renaissance poetry being yelled out by the Rexroth-Ferlinghetti troupe of minnesingers. Her big moment came when she plopped to the floor, realistic and natural; and one could hear the non-Whitmanesque Barbarians, sitting around with glasses of red wine, shriek out their prosody as they applauded the unconscious satire on the renaissance the gifted children were acting out. The children were healthy; but the poets were sick, crying for Mama, marihuana, tranquilizers, and the big tit they remembered about Mama.

One night, Caresse Crosby, the originator of the World Citizens Manifesto, dropped in at Rexroth's for a drink. Her pilot hero, Garry Davis, had made the movement and lost the movement in a know-nothing exhibition of confusion made doubly confusing. Since I had been a self-elected scootering citizen of the world, I appreciated the peculiar impact Davis had made a few years back, as he went passportless over borders, tearing up aerial frontiers, knocking over national barriers, wandering over Europe's postwar face like a wraith seeking international security against atom bombs. At the height of the movement, Caresse Crosby said, "We received over a half-million requests for membership, though we were hardly prepared for it. In fact, one never knew what Garry would do or say next. But he was the perfect symbol for us in his Air Force jacket and with that youthful, tow-headed look."

"What happened to the half-million letters?" I asked sympathetically. "Did you organize the letter writers into a One World Union?"

"We've only answered about 75,000 of the letters so far, unfortunately; and this is seven years after the events."

"Why bother now?" asked Rexroth. "You were sitting on something enormous but it's a dodo today.'

Later, some bulbous, bungling follower of Ezra Pound gave a reading at the Poetry Center, and another dodo bit the dust, with Rexroth and I deciding that the age was more Mongoloid and Mongrel than Insecure. Capitalism, American style, was quite secure, being more inventive and revolutionary than art and literature as it bumbled midway through the American Century; but the marginal mummers were the real conservative artificers, neither capable of revolutionary ideas nor conservative values.

I had many sentimental evenings with Rexroth and his wife, Martha; for Rexroth is the last of another epoch, a man with a facile mind, a great sotry-teller whose role as the *paterfamilias* of the children of the San Francisco renaissance was rather quixotic. I saw him in the role of a miniature Mencken, but more restrained, doing an eccentric oriental act at the Palace in New York.

We disemboweled an assortment of famous professors and their books, with Kenneth's sadness giving way to glee, especially as he turned his frenetic blade up one of Ezra Pound's many braying asses, who followed the Sage of Rapallo in all of his meanderings in the literary stables.

"He's a lousy translator, that man Pound, especially from the Chinese!" bellowed Rexroth as I went to sleep. That was not news about translators from the Chinese.

A few days later I was scootering again, going toward Hollywood on the wild coast road. State Highway One, which follows an almost primeval, mountainous terrain, is satisfying to the seeking eyes and the errant mind. It was silent, distant and august, with few cars or trucks zooming over it. The ocean, deep below, pounded on a wilderness of beaches left to themselves, uninhabited and tumultous. And I remembered an enforced walking trip that I had taken during the pre-Depression era, in 1928, on my way northeast from Hollywood. Being broke, I had walked almost the entire way from Hollywood and Vine to distant Seattle, with occasional lifts to help my wearying feet. I slept along the silent beaches, silent of human sounds, listening to the gigantic waves riot through the chilling nights; for my fire would go out, though I dragged the beaches for driftwood and piled my roaring fire high. I loved the driftwood, often refusing to burn pieces that nature had shaped and reshaped into natural sculpture in its Daliesque world, clockless and timeless, waiting for an artsy-craftsy aesthete to drift along and build himself a lamp. I saw the remains of ships that never came back, disintegrated and nameless, with spars, chains, booms and rusted anchors spewed up on the shores; and each night I created a fantastic history of men and ships as I slept on the beach and imagined the wrecking sequences that drove these ships on shoals. America and its once-great sailing ships became a nightly maritime masquerade in the fiery theater of burning driftwood as I lay and dreamed about our past, my past, and my future.

Then I was a wandering labor organizer; an itinerant labor journalist, with a few dozen bylines in labor papers. But I was broke, too, which was quite normal for me during the Depression; but I did manage to eat, having worked out an unbeatable if unpalatable diet that cost me only fifteen cents a day. Bread, American cheese and carrots contained all the dull values, especially iron. With iron in me, I was the tough organizer, talking industrial unionism, Trotskyism and poetry, in some peculiar yet related kinship, each to each. And now I remembered the events of 1928 as I scootered happily along the coast road, adrift in a natural state, feeling a sense of joy bending me toward the wild waves and the latter-day wilderness.

But the wilderness ended much before Monterey, at the shack of an old friend of those labor-conscious free-lancing days. Mr. W. M. was a science fiction writer, a serious one, with a whole host of Biblical asides to give his characters character. Apparently, science, or its fiction, could save the world if only the moral meanings of the Bible found their way into the metallic minds of the space cadets as they whirled between heaven and earth.

Mr. W. M. took a long time to write his stories, revamping and rewriting like Conrad, happy to have two hundred words completed each day. Wearing a black skullcap and faded dungarees, he sat amid three old typewriters, amid children crying for Mama (who was out working) and lived the life of science fiction. But each day had its earning power in a related way; for W. M. conducted a writing school, and the two spare typewriters were used by two students of his, who came late in the afternoon and sat alongside of him composing novels or science fiction. Later, over tea, Mr. W. M. criticized their works in progress and it began all over again the next day.

But he was a genteel, religious man, caught in a strange relationship to our real world, preferring his fiction and his skullcap as his total dress, along with his ancient dungarees; for obviously, everything else was a compromise.

It was eleven years since I had been to California and traveled the West, and my visual expectancies were mixed. For much of the United States is amazingly exciting as scenery, containing nature in its normal condition, when it is not despoiled by motels or signs that make each mountain and dale just another summit or valley advertising tourist buffoonery and emasculating the scenery. I had loved the West and had once known it intimately, having camped and lived on and off its earth on many occasions. But now I was a hesitant voyager in my own country, fearing to see over the hill, down the dale, across the valley and into the American dream. And what I often saw was a Dantean pit of commerce, spewing over in its inglorious cheapness and physical devastation, a merciless breeding ground for the shallow and the hollow, destroying man's eyes as well as nature's great gifts.

Between Carmel and Big Sur live America's maverick writers of prose and poetry; the independent men of letters, like Robinson Jeffers, whose poetry, with its Grecian embroidered tragedies, I had loved as a boy; and Henry Miller, who emblazoned the free libido on his skull, much as if it were an Anglo-Saxon heraldic symbol. But at forty-nine, I must have grown a great deal, since neither refurbished Greek tragedies nor Miller's excellencies as a man made me pause at Big Sur. I was stopped, however, by a landslide ten miles before Big Sur, and a dozen motorists and I dug our way out of the heaped-over road, using pieces of wood for shovels as we scraped away tons of earth that had spilled down after a big rain.

Only Spanish names remained of the Mexican and the Spanish past; with military reservations, and the once-lord of San Simeon, Hearst, having taken up the reserves of the much-past Californian frontier. These two symbols equated modern values—Hearst and arms, both having been the proving grounds for our age, now filled with old soldiers and film jungles. But

Hearst's land was going to the State of California; and a gigantic museum would replace San Simeon's muck-raking symbols. But California was more than that. It was America's fastest growing oasis. It was also Hollywood, where the marriage of the arts to the homosexual elite had taken root. Not strangely, the former fellow-traveling communists, in their distorted social dress, were the innovators of this trial marriage of values; once again the forerunners as destructive agents, making America shoddier to the mind.

I slept along the beaches, thinking my way back to the time when 1928 had a great rebellion in its heart; when each morning (and not solely because I was young) was a social, political and spiritual adventure. There was a vibrant energy in our thinking, in our rebelling, in fighting the status quo of the Depression-in-being as I hoboed over America, feeling her impact at every juncture and in every mile. We were another breed then, not yet "the careful, cautious, security-ridden generation now in our colleges," that J. Donald Adams, of the New York *Times* book review section, recently noted one Sunday morning.

And the beaches were not yet secured for the summer home ideal of progress; they were still wild, waiting for the real estate merchants to muck up and divide parcels of land; to build shoddy creations where things were better for everybody if left to their wild state. The new villages and progress-towns had not yet come, and one could scooter for a few miles without seeing a ranch-type house without a ranch in the background. Again I was in the world of the washing, roaring surf. Here I had burned many a steak black; and I remembered that in 1951 and 1952, Carl Sandburg used to walk up the four long flights to my apartment on Sixth Avenue, in New York, and ask, "What's for dinner, Harry?"

"Steak."

"Burn it black for me, like when we were hoboes."

I burned it black along the beaches, with the juices inside and the tang outside. I was the Chef de Cordon Hobo, a short-order cook from the Paris institution, reflecting now on things past, on the ways of poets when they take to the bush and the beaches for their emotional cuisine.

Three days later I was in Hollywood again, twenty-three years after my first youthful pilgrimage to the dead-heart of America. I had lunch with the magazine editor who had bought my offbeat stories, written as contrast to Hollywood's wooden world of make-believe. It was Mr. Z. who had gotten me halfway around the scootering world; for the ten articles and stories that he bought and published were guarantees against sudden dead-ends and enforced economic halts. Some of the stories I had concocted completely, especially one about houris in an invented Persian garden. For fifty American dollars, I wrote in this story, a man was welcomed to a blue-tiled garden, replete with private pools, bowers, delicate naked ladies—and he could have his pick of them. Knowing that some of my local readers might want to drop in, I told them that cameras were forbidden, since the garden was very intimate and photographs were thus taboo; that the place was frequented by amorous ambassadors and plenty of Teheran's plenipotentiaries; that if one were discreet, heaven would unfold.

Hundreds of letters came to the editor, many of them from American soldiers stationed in Teheran, who had not been able to find the Persian garden. They said it must be a fake. It was.

It was a fake as so much is in America, where exaggeration tends to become normal behavior. I had learned the art of concocting in the men's magazines after I had written a hundred pieces of pulsating pulp; where nothing worthy is confidential and everything useless is exposed; the sacred is profaned by association, and the profaned made sacred material; for without the latter the circulation of these magazines would disappear. Whatever was the fatal flaw in the American male, these magazines had found it. And, like so many others, I had several pseudonyms for the profaner magazines, though my purer work appeared under my own name. But with Hollywood looming up, I felt awed, for nothing I had concocted or jazzed-up could compare with Hollywood's cinematic horrors.

The editor spent a whole dollar on me in his welcoming

lunch; then he rushed back to his office. Still a novice to the ways and means of these magazines, I visited various departments of this libidinous factory. In the art section the airbrush got the nudes in the magazines through the mails, with the puritanical airbrush making Eve's pubic hairs rather fig-leaf-like —a solemn rite after the photographer got through with the buxom models posing for the assorted magazines Mr. Z. published. His best magazine, somewhat literary, came out on fine stock and with good photography, making the luscious models happy to be part of American culture. His worst magazines lacked everything, though they had lots of lewd nudes splashed across their pulpy pages. When I saw stories by William Saroyan as well as by serious poets I had once known, I was convinced that this was the way of the poet's world and that I had not strayed too far off the literary path of righteousness.

But if this was not the gutter of publishing, it was just an inch above it, or near enough to the sewer so that it could be mistaken for garbage. Modestly, however, I wrote pieces that seldom made me ashamed of myself. It was the photos and the suggestive leads, not the ingredients of words and situations, that made so many of these magazines offensive and hardly Rabelaisian, which their editors always insisted they were. And if I appear to be biting the hands that fed me, I am biting manicured claws.

One day, while scooter-sightseeing around Hollywood's sagging sights, I ran through a red light. A motorcycle cop, noting my foreign plate, pulled me to the side.

"Where are you from?"

"I've just come from Australia," I said, hoping that our excellent relations with the Aussies would see me through this.

"And that plate? Is that Australian, too?"

"No, it's an Italian international plate."

"Can I see your driver's license?" asked the cop.

"You don't need one for a scooter in Australia," I said, suddenly realizing that I had driven around the world without a driver's license, and to get a ticket now would spoil something. After all, I had passed red lights in a lot of countries, but a totalitarian cop in Yugoslavia had given me my only traffic fine.

"Well, you need a license in this country. Are you Italian or Australian?"

"I'm American," I said, adding to the international confusion.

The cop, now helpless, said that I was in for a lot of serious trouble. He handed me a copy of the Vehicle Code of California, with instructions to go for a driver's examination the next day. But I wanted to remain in that "serious trouble" state and drive across the United States, totally licenseless before life in New York, so well licensed, enveloped me again. In leaving, I remarked humorously that I had driven around the world because I was not sure that I could make the necessary U turn needed to get a license; that I had taken the world at its widest U arc as it was the only way I could turn around and get back to New York. But the cop was not amused and my U joke fell wide of its mark.

The next day, however, I ignored the summons to take the driving test. Instead, I drove to Los Angeles and received a handsome check from Frank Howe, a Mobilgas publicity executive who was using some of my photos for their house organ. An Indian journalist named Padmanhaba, who was finishing a five-year training period on the *Los Angeles Times,* gave me a politically-minded interview, with the lead, "Plight of World Dismays Writer . . . eleven-nation tour by motor scooter reveals universal trouble, traveler reports." Mr. Padmanhaba wrote a thousand-word interview that had guts and grace in it, and quoted me as saying that only a United States of Free Asia could end the communist threat of Khrushchev and his Marxist hordes. About religion in Asia, I was quoted as saying, "The leaders and people of Asia must realize that they cannot think of religion in terms of Biblical times. Science and man have met and fused and religious rivalries must come to terms with the twentieth century."

I had come to terms with myself as fast as I could travel

in myself. I had experienced everything human, so that nothing was alien to me, with the massive notes of my journal trailing behind me like a Chinese kite of many tails. Now I must sit down and put the bulking data in order before I was overcome by its size. I no longer kept notes; for I was in a mental miasma, incapable of further sensitive impressions and registrations. My camera eye was bloodshot, and all was fastly becoming black and white in a gyrating flotsam of active and static observations. But now, about to start across the wide roads of heavily trafficked United States, my eyes, I soon discovered, were almost out of focus. Too late, I realized that the Persian and Indian sun, where I seldom wore dark glasses, had brought this foggy state to my vision. An eye doctor, whom I saw socially, told me to stay off the roads, not to drive, not to read, and not to use my eyes for anything but an occasional proper glance at pretty girls. I began to fear blindness; for my mother had gone blind in her old age, and we both had the same deep blue eyes.

I read poetry, instead, accepting an invitation from the literary critic, Lawrence Lipton, who had invited fifty people to his home in Venice. William Saroyan, who lived in nearby Malibu, happily turned up, too, for this eccentric evening of poetry and tape-recordings; for it was Hollywood's renaissance competing with the San Francisco's renaissance in a splutter and splurge of fantasy and reality. But before poetry overwhelmed us, Saroyan apologized for not answering my letter regarding Dr. Hovanissian, the Persian doctor who had saved my life in Omar Khayyam's Nishapur. "You had all the Armenians in the United States helping him out, including Dr. John Hagopian and Reuban Mamoulian, and I'm sure Dr. Hovanissian is already here, practicing on Sunset Boulevard." But Dr. Hovanissian, I learned later, had not arrived.

Saroyan, Biblical-minded, read Genesis like a Billy Graham turned truck driver, with Lipton taking it down on tape, as if this were the beginning of Creation all over again. But if Saroyan did not create the world in one evening of reading, his audience believed that he was capable of remaking it, prose-wise, with whimsey, myths and folky fiction. He had just sold another story to Mr. Z., my security-man, who had taken me to a dollar's worth of lunch; but Genesis was pure creation, with only Noah and the Ark for Saroyan's Biblical backdrop.

I read a long poem called "The Return" from my new book, *My Father's Time:* an emotional Spoon River time-piece; a portrait of the lower East Side of New York up against itself and my childhood. And since I was to leave for New York in a few days, I was completely subjective as I saw the tatters of my past raging along the windy docks of South Street, where my family had once lived. I had worked on the poem for almost a year, as if to pay myself back for my neglect. It was an indictment of my private shambles; of the doomed, domed world before World War I, when the ten nationalities on my block were trying to become American in a hurry. Only too conscious of my present, "The Return" distilled my past, like this:

> Go with me and be my memory
> And we will all our terrors find;
> The equal desolations of our route
> Will shape murderous models of our minds;
> In some private purgatory of our race
> You'll find your image and your sacred place,
> For the Century stammers with disorders
> And every factor animates its actors,
> And like a public servant with a private purpose
> Your morality is your own security;
>   The void's within, but outside
>   The City invests its suicide.
>
> It rises within your holier family
> Dueling in a bedroom's tightened air;
> Paternal, real, consummate with love,
> You recall a hundred taut realities
> Mothering your infancy and savagery;
> And as you face the things you were,
> The lives you lived and the lives you lied
> In childhood's green magic and black misery—
> You re-enact the stages of all your raging ages.
>
> Indelible, nimble, acquainted and distorted,
> Our youth inherits many modern maladies,
> Heroic external whims and mystical fallacies.

Though the world is young, we are old,
Abysmal, indexed, having secret passports
To the sacred fountains of ourselves,
To water the places that sanctify
The massing mummeries of our eye.

The street of my childhood is a jewelled slum
Festooning the awkward mesmerisms of innocence.
Yet the picture here is all wrong;
It was a street of murder, mimicry and malice.
In my youth it killed
Ten citizens of fear and poverty . . .

This street is a tomb, for I saw it yesterday:
A soundless crumbling church
Near a pitiful synagogue
Facing two exits to vision and piety.
Tattered newspapers, in the cornered garbage,
Vie with the names of streets I ravaged
To lighten my burdens as a boy.
Their past stab my dust-laden eyes,
As antennaed, pigeonless roofs impale me,
Corrupting the low skies
With ambushes of grief.

The ghosting docks I knew are still with ships,
Loading every commodity of our industry;
And massive buildings have shimmered heavenward,
To make new tenements when the Century ends;
I know, it was like that forty years ago.
I see similar faces cradled to their mothers.

> The house that I was born in
> Is now a parking lot for trucks;
> Another, boarded and unevenly shuttered,
> Closes in its evil with the night;
> And the factory where I stole candy
> On Armistice Day, now celebrates decay.

Some old citizens are still here, waiting
For time's induction to another stillness.
The Negroes have come and made a church
And their singing is strangely real to me.
The fat little stores, so ancient, then,

Wear their age with commercial indifference;
And on some suburb's green patch, the friends
I had as a boy, have settled
Their middle age's brooding disenchantment.

I walk alone, vainly competing with my sentiments,
Remembering what I was and what I have become,
Breathing from each street a fable of my world.
I grasp for the age that turned my Century
To create my innocence and my rage.
I remember the strong boys, who robed their fists
As boxers, to general us in gangs;
And those who became thieves, and murdered, later,
Incorporated into crime like movie-scripts . . .
    I wear ten scars for cruelty, five for cowardice;
    I wear three for fear, two for innocence,
    And one for my father's rage.

And I remember those who walked alone
To listen to themselves as they paced careers;
The teacher who lectured to me avidly
On the Russian and the German souls;
But he is dead, like the nations
He had not known too well.

This is a moment in another Hell,
And I know each step that walked my way.
In some blackness before my front door
I saw ten nationalities become American.
Brimming with venom, filled with Europe,
And cursing their strangeness, I knew
Ten versions of malice, race and hate . . .
And I knew their vagrant daughters
Pregnant with bastards at fifteen . . .

A young artist named Saul White did an impressionistic sketch of the poem, with a church and a synagogue, buttressed by large, penetrating eyes. He put trucks, ashcans and the bric-a-brac of the tenement world into his "visionary" drawing. I was, of course, flattered by the drawing, which he presented to me as a gift while Lipton was reading a poem on the birth of jazz. Lipton, an energetic man and writer of many parts, and once the husband of detective-story writer Craig Rice, finally summed up his own "renaissance," by saying, "Rexroth is overdoing it with saxes and stuff. Poetry needs drums first, 'cause you've got to sound out the sound-values before you start orchestrating it fully for more instruments. We'll try drums here for a year, then gradually we'll add other instruments to build up the background. But it is poetry first before it is anything else—so listen to this."

We heard a tape recording of an evening given over to poems and drums. If this was new, it was only new to Lipton. I had heard drum-poetry during the twenties in Greenwich Village and Harlem cellar-clubs, along with speakeasy music, when poetry was not a muted sound in the orchestrations of Prohibition. But in Hollywood, or in nearby Venice, it was considered new when two people said so—and Lipton was at least two people saying so.

Adding to our audience was a pretty girl, Iris Adams, who had come to her first reading of poetry. She was the daughter of Craig Rice (by a former marriage), and she scoffed like an amateur detective at the proceedings. "And you don't have long hair, either; nor are you crew-cut, exactly. God, what odd people they were!" she said, when I took her home. "But Saroyan sounded real *real*." Lipton's renaissance people were dungareed and pedigreed bohemians, arty and farty, normal and abnormal, using pod and other drug-type stimulants to stimulate their life of art. And whatever the lovely lady thought, I thought, having seen enough of the dungareed life of art to last the rest of my literary safari.

I was back on the scooter, beginning the long haul home on March fourth, on America's open roads to adventure, sudden death, and assorted forms of scootering mayhem. Victorville, California, was my first night-stop before crossing the now-cooled Mojave desert, which was like a spring song before the bubbling summer heat inflamed it into fiery pits. I had waited out the winter, or so I thought, as I entered the petrified forests and the painted deserts, where the Old West lingers in a hundred trading posts, each selling trinkets of America's Western epoch now gone stale and commercial, through which I rode like a demon on two wheels, unable to take in any more vistas and visions.

I saw things as a sleep-walker in surface nightmares as I passed by lava beds, soaring peaks, crater lakes, huge dams and little orange groves. There were race tracks, missions, pueblos, and Indians without ceremonials filling in the American Heritage that I was going back to. I had gone over this land many times before, living in the hobo jungles, happy to get my share of the mulligan stew the hoboes cooked up in tin cans. But the hobo jungles had gone into the heritage, along with the hoboes, who were likely eating tranquilizer pills, waiting for the green light to security, not the red-ball freight train pounding down the American tracks. It was part of the world gone over the hill, momentarily recaptured. But, then, everything old was gone; for the frontier had vanished quickly and indecently. All the adventurers and their free-wheeling ways, all the early fancies and the later poetry, had gone into the dead-end end. And all that I saw were well-placed, streamlined roads, racing to the American horizon and back again. I was back-tracking into the empty land, yet too full; into the scene butchered and the scene beautiful, where our people were facing the end-all of American progress and its processions. In one hundred and seventy-five years we had made a world of joy and plentitude; but whether it was good or not, I could no longer tell. It was called progress.

Ninety-seven per cent of American land is given over to cities, towns, villages, farms, roads and industrial enterprises, and only three per cent is natural "wilderness" and national parks. There is no room any more except upwards or downwards, and therefore the frontier is something quite holy in our Western-ridden consciousness, sacred but gone, except in our sad memories.

I reached Albuquerque in four tense riding days, sleeping over at Victorville, Kingman and Holbrook in one-night functional motels. The roads, when they were four-laned, were not scooter hazards; but on the narrow, two-lane roads, the gigantic trucks would lift me from the ground. When the winds rose, they lashed into my windscreen, careening me across the trafficked roads like the furies gone madder. My nerves were shot; for I was no longer the carefree adventurer of Greece, Turkey and Iran, where road traffic did not exist any more than roads did. Oddly, I had dreamed of American roads, thinking how smooth would be my passage to New York. But it soon became obvious that it was suicidal; that double-trailer trucks were not covered wagons with singing Turkoman tribesmen. The trailers were murderous, massive, engined chariots carrying American goods; and a man on a scooter was an innocent stranger riding a strange thing. To keep from being blown off the road and killed, I worked out a safety factor whenever these zooming trucks approached me. I braked for a quick moment, a moment that had me suspended between sudden death, via sucked-in collision, or being thrown off by gusts generated by the sudden vacuum. Before I reached New York, to pass the time of travail, I counted 9,741 trucks roaring by with the thrusts of jets. Occasionally, when weary with braking and this madness, I would try and wave a truck down to a slower passing pace; but I must have looked like a madman as I semaphored with my right hand, my left clinging angrily to the handle bar.

Albuquerque had changed considerably since the genial cooks of the Harvey House had me wash Depression-day dishes for a meal. The cooks had fed me, given me extra sandwiches for the road, and told me to head east, not west. Heading west meant running into John Steinbeck's *Grapes of Wrath* upon the land, and more hunger than there was in the East. But Albuquerque was another place and another time now. It was modern adobe, with land-shy easterners, who had become westerners, fencing themselves in. Every adobe house had a fencing-wall to keep intact a vision of middle-class western splendor. But it was a splendid land, where a man could live and a man could die. Here a man could write and live widely, hugely and earthily if he had the cash and not the peculiar American poverty of the spirit. For in real poverty there are few places to inhabit in this world. In Iran, one can live, and one does, on tradition; but in New Mexico one can't and one does not, no matter whether one is American, Mexican or Indian. It is a nomad land without nomads, rich with colors and intimate flavors of basic living; like the Turkoman's world, with its deserts and mountains.

I stayed with Bill and Maxine Jackson, Australian acquaintances, who had settled in Albuquerque. Bill was a petroleum engineer who had oiled his way through the embattled Middle East oil fields until the certitude of oil, plus politics as a way of life, became too uncertain. Now he was an American citizen, aware that his birthplace was something vague in a vaguer past; but it was good to hear his Australian accent, still very much present; to listen to his Aussie slang, which he produced when he saw that I craved for the sharp flavors of "My bloody oath!" as we settled down to a drink and American maps.

"Get off U.S. Sixty-six, Harry. You'll be hitting snow, tornadoes, twisters, cyclones, and just ordinary one-hundred-mile winds," he said. "Besides, the scenery is much better down under on Route Sixty, which I ride regularly to see my bloody clients. But you'll get lots of bloody trucks either way; that is, unless you take the bloody side-roads all the way to New York."

I was tempted to go via the side-roads, but it would take much longer than the twelve days I had given myself to scoot across to New York. My three-day stopover in Bill's Little Australia in Albuquerque became a progress report as I told handsome Bill and Undine-like Maxine, who was writing short stories, how much Australia had developed since Bill had left once-backward Terra Australis. I was always, emotionally, pro-Australian, but in an amateur way, hardly the professional propagandist, though I had written dozens of articles and stories about nature's land, Down Under, to make friends believe that I was in it for the shillings and the pence. Besides, as an ex-athlete, I acknowledged the superiority of their swimmers and tennis players, if writing on sports can be considered propaganda. They were clumsy at baseball, and probably could not beat the agile

Japanese at it; but the Australians could take all prizes for knowing how to relax. Also, psychoanalysts in Australia were still patient-poor and not money-rich; and couches were still something the Australians used for a "nappo" and not for delivering the conscious mind up to an unconscious middleman of the psyche, standing over your comatose soul like a soldier at parade rest.

Rested, I resumed my journey, with nightly stopovers at Fort Sumner, Lubbock and Dallas, lost in the Texas winds, in the flat areas of geographical disenchantments, my eyes a red-rimmed nest of double vision. Trucks, trucks, and more trucks shattered and thundered as I went over the plains and the hills, passing the grave of Billy the Kid. A friend, Charles Neider, had just published a novel about Billy—and so I sent Neider a postcard of the fenced-off grave in back of an Old West museum which was in back of a weary grocery store. Billy had killed and been killed. Billy had made another part of the American folklore, one that was populated with frontier gangsters and gunmen. Billy was ten miles off the main road, amid pasturing goats and running dogs. My friend Charles Neider's Billy was behind a fence and behind the American Dream. The gunman of the murderous impulse had helped to make violence a national characteristic. The Killer was still the abysmal American Hero, and for fifty cents I saw old photographs of the Kid, old guns and artifacts about death in the once-Wild West.

I rode on, conscious of the violence fixed to this endless treadmill of truck traffic. The last wonderful pleasures of scootering were left behind in Albuquerque, with elusive mental devils telling me that the road back was the road home to unconscious suicide.

To continue in my state of fatigue meant eventual collision. Winter was still hugging the land, and I was gripped by the same fatalism that I had experienced in Ceylon before I was sideswiped by the bus. Yet every day became 250 wearying miles; then cheap motels, where truck drivers lay over for the night unaware that I was their strange enemy, a quixotic one who went to sleep dreaming of slaughtering them en masse before they killed me. For death was on the curves along the sodden roads, deep in the traffic lanes and the wintry winds, as I scootered on the rained-out, skidding, oozing, American earth, in the hundreds of ways an exhausted traveler imagines it before reality presents him with still another variant.

I write of this at length because there was nothing else in my mind, then attuned to the pandemonium of presentiments and physical fatigue. The Y.M.C.A.s and the sleazy motels I stopped at added to the morbid facts and fantasies as I rode through Texarkana, Little Rock, Atlanta, and Gaffney, North Carolina. I was closer to winter when I reached South Boston, Virginia; a state so beautiful yet so racist, with the tobacco tang breezing into my windscreen. I rode to Richmond in a cold dull winter rain, remembering the warm monsoons of India, for twenty-four hours of unending driving—then Washington, D. C., deep in a private coma, unfeeling and all rained on. I was wet and cold, wearing a ski sweater and an old trench coat, frozen into my follies, unable to make contact with myself, lost in the limbo, in pain, fear, roads, trucks and America. I was half-blind and shabby, wondering when the fatal skid would come; the last skid on the seas of macadam as I entered Washington for a late dinner, before continuing through the night to New York.

I had missed the presidential election while at Melbourne's Olympic athletic carnival, but I had not missed its political circus, which was reported adequately enough. The gracious Hamlet, Adlai Stevenson, had once again defeated himself. And as I ate my last supper as a traveler, I amused myself, to get away from my No-Doze coma, with political conjectures and social analysis. It helped to decrease my scootering tensions, now brought to the dramatic pitch of the last act, last curtain scene. I thought, with due exaggeration, that we were now the leaders of many tattered international armies; that our national ego had been shattered and scattered as we drifted within the Khrushchev-Nasser orbit-of-no-return; that our moral strength had seeped away despite our moral indignation regarding the uprisings in Poland, East Germany and Hungary. We had learned the arduous duplicity of diplomatic talk and had become quite expert at placating the totalitarian enemy; we learned that we would do nothing about these risings except talk; and that our possible victories against totalitarian oppression were cheaply given away every time we were on the verge of winning something tangible in this permanent war. And though we had literally founded the public relations industry, our relations with our democratic allies were now neatly confusing. We had become anemic and petulant instead of politically mature; and when the towering United Nations structure, to which we had confined all our possible victories as well as our future, caved in, we would be neatly buried there by our total naïvete, if not by the endless but terrible errors we had made. But there was always Billy Graham ready to touch off another spiritual revival; for he was quoted in the Washington paper that I read over my steak as saying, "So much of American advertising is a lie!"

Were the "lies" of the poet and the "lies" of the politician, via Madison Avenue, really involved in the American Dream? I did not attempt an answer in Washington over the second cup of coffee I swallowed, readying myself for the road again. For I was soon wallowing in the night fog and the rain embracing Washington, the moral frontier of America.

I had, I reflected through the night, not been the best ambassador of our purest puritanical morality. But I was by nature a decent enough citizen—and the long night's ride through Washington, Philadelphia, Newark, watching the darkness turn into morning, also turned the last ten miles into comedy. For late the next morning, my scooter stopped dead ten miles from the George Washington Bridge. The rear wheel, locked into the worn-out rear brake, would no longer turn over. In an hour I was to meet the press on the New York side. I phoned the Vespa Company, and a Mr. Montizemolo, a dapper Italian ex-marquis turned noble salesman of the waspish-shaped scooter, said that he would get me out of the dramatic dilemma and send a mechanic to unfreeze the brake. But I could not wait, knowing the prompt press, which had been alerted by some public-relations Madison Avenue handout. Impatient to get it over with, I hired a truck to carry the scooter and me to the New York side. If I got there before the press, I could save face.

Three miles from the bridge I met a flaming red scooter hurtling down the road. The comedy was over when George Nicholson, the "mechanic," got me on the road again. I went, sans brakes, over the latticed bridge, into the New York skyline that I had waved away fifteen months earlier, home from my odyssey on two wheels.

I had solved some things during the journey into myself and into the world. The subjective and the objective, in their separations, could be partially summarized. The shades of meaning and their emotional areas were not places of a physical geography, but a private map of my internal world. I had won a little victory over myself and discovered someone called Harry Roskolenko, F.O.B., in the long passage from maps and conflicts into myself. A vast array of facts and fictions had helped to sharpen the process of self-discovery as I discovered others, less attuned in 1957, for any sort of a world.

I had twenty-four dollars left on the day I arrived; but a friend told me that a book of mine had earned seventeen thousand dollars in *possible* royalties. The paperback edition had sold 750,000 copies; but all that I got from it was one hundred and fifty dollars for the American rights, received before I left for Paris, on January 3, 1956. If this was ironical, I was making the least of it. But as Neitzsche had said: "And which of us poets hath not adulterated his wine?" I had, too often. And who hath not adulterated his basic values? I had, too often. But gas and oil can be reckoned up, if other things are still within the intangibles of my journey. I had spent ninety-seven dollars on fuel. I had traveled 37,000 miles, 21,000 of them by scooter, on the folkways and byways of the world. The pilgrimage was done, the free-fancies completed; but the dreams and the challenges that some men have I had known intimately.

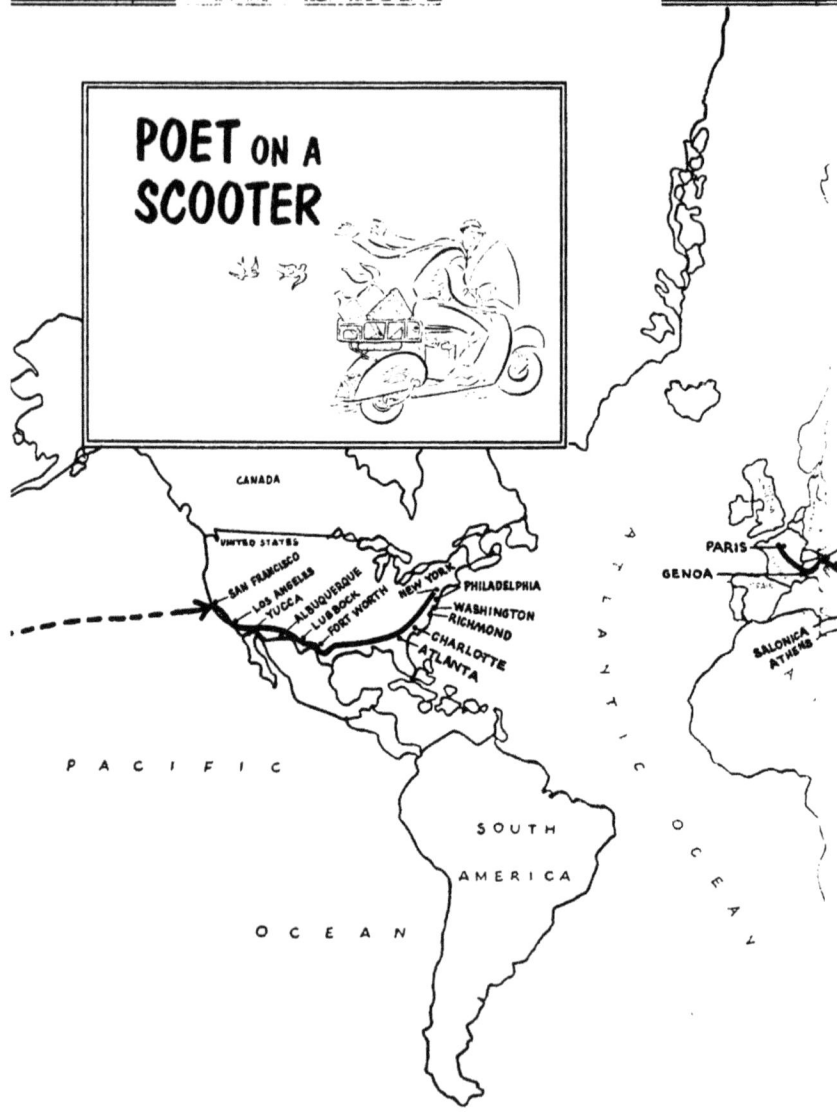

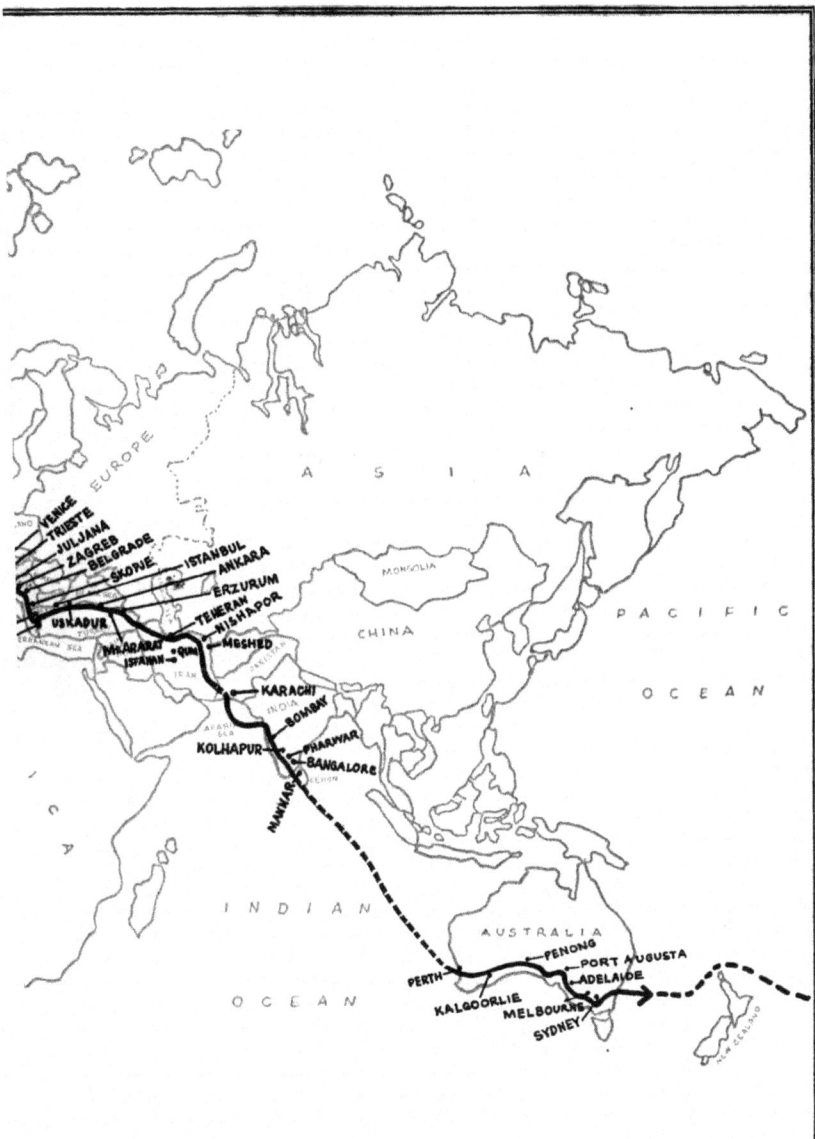

www.ingramcontent.com/pod-product-compliance
Ingram Content Group UK Ltd.
Pitfield, Milton Keynes, MK11 3LW, UK
UKHW021342051225
9402UKWH00045B/822